THE INTIMATE COMMODITY

• • • • • • • •

THE INTIMATE COMMODITY

Food and the Development of the Agro-Industrial Complex in Canada

•••••••

ANTHONY WINSON

Department of Sociology and Anthropology
University of Guelph

GARAMOND PRESS

Publisher: Peter Saunders
Editor: Robert Clarke
Typesetting and design: Robin Brass Studio
Printed and bound in Canada

Canadian Cataloguing in Publication Data

Winson, Anthony, 1952–

   The intimate commodity : food and the development of the agro-industrial complex in Canada

ISBN 0-920059-19-8

1. Food industry and trade – Canada. I. Title

HD9014.C32W56 1993   338.1'971   C93-094073-3

The publication of this book has been supported by a grant from the Canadian Studies Funding Program of the Department of the Secretary of State. The opinions expressed do not necessarily reflect the views of the Government of Canada.

CONTENTS

• • • • • • • •

• • • • •

PART I

THE PRODUCERS AND THEIR STRUGGLE FOR POWER

CHAPTER ONE

CHAPTER TWO

CHAPTER THREE

CHAPTER SEVEN

FOOD RETAILERS: THE NEW MASTERS OF THE FOOD SYSTEM   160

CHAPTER EIGHT

RESTRUCTURING AND CRISIS IN THE CANADIAN FOOD SYSTEM   184
Sandra Watson and Anthony Winson

• • • • •

LIST OF TABLES

LIST OF FIGURES

PREFACE

• • • • • • • •

This book is about the complex of economic, social and political practices linked to the production of food in Canadian society. The subject matter of the book has a relatively long history. It is also notoriously complex and under-researched. As a result, it would be misleading to claim that the present work is anything more than an introduction to the subject. A truly comprehensive treatment would have required a very long book indeed and would have defeated my original purpose: to give readers some general insights into basic but important issues, such as how our food system is organized, how it grew to be that way, and how to critically assess the consequences of existing arrangements for the various social groups affected.

Much of what has been written on what I broadly refer to as the Canadian food system is restricted to agriculture, and to some extent also deals with the matter of rural community. We do not as yet have a body of research- as do the French and the Americans for their reality- that provides a complete picture of the wider Canadian food economy, of which agriculture is but a part. I hope this book will contribute to this task – one that has, fortunately, been taken up by a few other writers as well.

In the course of working on this book and conducting the fieldwork upon which some of it is based, I had the benefit of considerable assistance, time, and encouragement on the part of a sizable number of people. We interviewed many farm operators and their spouses in southern Ontario and Nova Scotia during the course of our fieldwork, and we owe them a debt of gratitude for the generosity they invariably showed us in our search for information and understanding. Valda Gillis and Beverly Wemp provided invaluable service with the on-farm interviews. We owe a similar debt to the individuals I interviewed in co-operative and private food- processing firms in both provinces. I am also grateful for funding received from the Social Science and

Humanities Research Council of Canada for much of the early fieldwork, and also for funding provided by the University of Guelph.

I would like to thank George McLaughlin, Brian Kipping, Quentin Chiotti, and Phil Cairns for information about the origins and current structure of the Ontario Milk Marketing Board, Sue Cox for insights into the politics of food banks, Hon. Ralph Ferguson and Nick Heisler for up-to-date information on food commodity prices and the retail food industry, John Mumford for information on the fruit and vegetable industry, Murray Knuttila for important material on E.A. Partridge, Ted Chudleigh for insights into the Ontario food-processing industry, Kip Connolly and Janet Dasenger for assistance with interviews with food industry plant workers, and William McLean for details of the early development of Canada Packers and the role of his father, J.S. McLean, in building up that firm.

A number of people have provided inspiration and encouragement for this project. Alex Sim's progressive agrarian values and deep-seated concern for the fate of rural communities have certainly been one inspiration. Bob Stirling provided encouragement and wise criticism when it was needed. At the University of Guelph I am grateful for the support of Jim Mahone and for helpful discussions with Hans Bakker, Stan Barrett, Ken Menzies, and numerous other friends and colleagues. I owe much as well to Peter Saunders, my publisher, for his sustained interest in the idea behind this book. Robert Clarke must also be highly commended for his diligence and painstaking efforts in the role of editor.

This book would not have been possible without the support provided by my family- Dorothy, Devin, Mark, and Angela. Ruth Lesins's ongoing co-operation has been especially helpful. During the last two years or so the continuing reassurance and emotional support of Barbara Webster-Powell have been very important to me. Finally, Sandra Watson helped in innumerable ways during the later phase of this project and is largely responsible for the telling and poignant testimonies of unemployed food-industry workers that have been incorporated into chapter eight. I am grateful for her enthusiasm, persistence, and tolerance.

A final debt is owed to Ellen Nilsen (1963-1990) for her love and encouragement when this project was initially taking shape. This book is dedicated to her memory, and to that of her mother and my good friend, Mona Nilsen (1923-1989).

ANTHONY WINSON
Guelph, July 1993

INTRODUCTION:
PRODUCTION,
CONSUMPTION, AND POWER
IN THE FOOD ECONOMY

•••••••

In the twentieth century the control of our foodways has been steadily taken over by large private corporations. As a result, as anthropologist Marvin Harris notes, it has become increasingly true that what is good to eat is good to sell (1985:248).

Certainly, within capitalist societies food has become a commodity – one in a long line of goods and services produced and sold. This means that the production, transportation, distribution, and consumption of food are subject to the same fundamental social forces and economic "laws" that work on and apply to other commodities. But it is also clear that this commodity, food, has its own very specific and significant socio-economic, cultural, and political characteristics. These peculiarities involve not only how food is consumed, but also how it is produced.

To begin with, food has a distinct relationship to social experience. Many other commodities today seem to fragment our social life: examples are the socially disintegrative effects of television or portable Walkman-style radios or cassette players. The automobile, perhaps the central commodity of the North American lifestyle, seems to isolate each one of us more and more from our fellow human beings as it becomes the primary means of moving human bodies from one point to another on a regular basis. The same argument can be made, with some modifications, about the impact of many of the other

1

major commodities that help to structure our lives in contemporary capitalist society, from houses to Nintendo games. But the rituals of food preparation and communal food consumption have played a central, integrative role in human society. They have formed an essential means of bringing people to-gether – of establishing human existence as a *social* existence.

Food: The Dimension of Consumption

As Peter Farb and George Armelagos indicate, the English word "compan-ion" is derived from French and Latin words that mean "one who eats bread with another" (1980:4). Farb and Armelagos also point out that the eating cus-toms of the Chinese are tightly interwoven with social transactions: "The giv-ing and sharing of food is the prototypic relationship in Chinese society, as if the word were literally made flesh.... No important business transaction and no marriage arrangement are ever concluded without the sharing of food. The quality of the meal and its setting convey a more subtle social message than anything that is consciously verbalized; attitudes that would be impolite if stated directly are communicated through the food channel" (ibid., 5). For the Malays of Southeast Asia, food – or more specifically a certain kind of food, rice – is thought to possess an essential life force. In Malay culture the cere-monies that mark the basic stages of life – birth, coming of age, marriage, and death – all involve a symbolic meal of rice. For other cultures, corn or maize plays much the same role. As one student of the Pueblo Indians of the south-western United States said, "When I take away corn from such people, I take away not only nutrition, not just a loved food. I take away an entire life and the meaning of life" (quoted in Farb and Armelagos, 1980:7).

Over the last few decades, family life in Western capitalist societies has become more and more disjointed, and the integrative function of the daily family meal has lost significance because of basic economic pressures that have forced parents – women as well as men – out of the home and into the labour market. Even so, for many people the dramatic growth in meals eaten away from the home signifies the displacement of a home-based integrative occasion – the family meal – to another locale, rather than the complete disap-pearance of the role of food in bringing people together.

I believe it is possible to extend this thesis concerning the socio-cultural sig-nificance of food even further, to say that food has a special role in the reproduc-tion of our species. The consumption of food and drink often provides the con-text for the initiation and development of romantic and sexual relationships. Western medieval lore is replete with references attesting to the relationship between food and sex. As Madeleine Pelner Cosman writes in her book on medieval feasts and ceremony, many wines and foods were believed to promote sexual desire. She notes that Chaucer's *Canterbury Tales* has numerous references to food that allude to its aphrodisiac qualities, such as the reference of a lover attempting to encourage the sexual desires of a lady by sending her fine mead, spiced ale, "piment," and wafers piping hot (Cosman, 1976:109). St. Thomas Aquinas argued that "gluttony and lust are concerned with pleasure of touch in matters of food and sex." Accordingly, an increasingly puritanical church was moved to forbid the partaking of those foods most pleasing to the palate, because of its belief that they were the strongest incentive to lust (ibid.).

Today, it seems we are rediscovering the importance of food for our social and psycho-sexual lives. The immensely popular film *Babette's Feast*, set in the 19th century, provides a particularly good example of the role of cooking and eating in creating a communal dimension to our lives. Laura Esquivel's best-selling novel *Like Water for Chocolate*, a classic love story set in turn-of-the-century Mexico, uses the preparation and consumption of food as a recurring catalyst that moves events forward. In Esquivel's work food, love, and lust are always intertwined, as when Tita, the leading character of the novel, concocts an exquisite meal that includes the petals from roses given to her by her lover. The meal has a dramatic effect on everyone present. "It was as if a strange alchemical process had disolved her entire being in the rose petal sauce, in the tender flesh of the quails, in the wine, in every one of the meal's aromas. That was the way she entered Pedro's body, hot, voluptuous, perfumed, totally sensuous" (1990:52).

In modern times, of course, most of the ceremony linking food and sex has been lost. However, even now the "candlelight dinner" is a central motif in the emotional and sexual life of men and women in contemporary Western societies.

If food and sexuality have a long history together, so too do food and spirituality. Cosman argues that eating habits in medieval times determined a human being's association with the pantheon of virtues and vices:

Gula, gluttony, that seducing sin which had corrupted the world, had a medieval definition both fascinating and unsettling. Just as Adam ate his way out of Paradise, so man eats his way into sin. Surprisingly, the dire and deadly sin to which writers such as Chaucer and a host of theologians ascribed Adam's loss of Eden was not pride but gluttony.... While other of the Seven Deadly Sins sometimes led the way to Hell, gluttony was the most simplistic explanation for sin in this world and man's temptation to it. Conversely, it was the most complex: the sacramental act of eating the body and blood of Christ had its perfect parallel, its special urgency, and its graphic vividness, as an undoing, by eating, of an evil deed caused by eating. As Adam ate his way to sin, so man might eat his way to salvation. (1976:116-20)

If food and drink have become commodities, they are "intimate" commodities like perhaps no other. In the process of their consumption we take them inside our very bodies, a fact that gives them a special significance denied such "externally" consumed commodities as refrigerators, automobiles, house paint, or television sets. Moreover, unlike so many other goods that we produce and consume in capitalist society, food is an *essential commodity*: we literally cannot live without it (although this is not to say that all of the processed food products for sale today are essential). While it may be a major inconvenience to be deprived of a car, to be deprived of food has consequences that are nothing less than catastrophic. It is not surprising, then, to find that when people have faced food shortages and hunger as children, much later in life they still tend to have strong views about the wasting of food: the generation of Canadians who lived through the Great Depression of the 1930s is a case in point. Societies that have faced life-threatening shortages of food – such as, for example, a number of European countries during the Second World War – have often later enacted legislation to ensure adequate supplies of foodstuffs. This kind of legislation has sometimes shaped the entire agricultural policy of a nation.[1] Even food in abundance but of insufficient quality can have serious consequences – far removed from the rather more meagre consequences of being deprived of colour television and having to make do with a black and white set.

Today the problems of the sufficiency, quality, and safety of food vary enormously in significance according to the social class, and the society, people belong to. For a sizable proportion of the population in the rich capitalist societies, the problem of achieving a sufficient caloric intake on a daily basis has been overtaken by the relatively new problem – in historical terms – of restricting daily caloric intake, and especially of reducing the consumption of fat-rich foods associated with cardiovascular disease (OECD, 1979). Yet within these same societies, some people – a minority but still far too large a number to be ignored – have been forced to rely on food banks and other sources of food donations simply to meet basic needs. Within this group, children figure prominently. Canada is no exception to this contradictory reality, as a recent study by the Edmonton Food Policy Council has demonstrated. In that city in 1991, of the one in five

households considered to be "low income," one-half of the families reported
that they could not afford to give their children good quality food or a variety of
food. Almost 20 per cent had cut down on the size of their children's meals because of a lack of food, and 15 per cent of these households said their children
were going hungry. The level of hunger reported by adults in these families was
considerably higher than for children, with over one-half of the adults from low-
income families reporting a state of hunger. This suggests that "parents deprive
themselves in order to provide food for their children."[2]

Social inequality and food – the quantity, quality, and type of food consumed,
and the style of its consumption – have long been closely related. Ceremonies
centred around the consumption of food once served to inhibit the development of social inequalities that would have otherwise been produced by the
unequal distribution of the natural resources available to different members of a
group. Competitive feasting is found in numerous societies as a mechanism to
redistribute resources in a more efficient manner (Farb and Armelagos,
1980:151), with the potlatch ceremony of the West Coast indigenous population
of North America one of the better known examples. More typically, however,
food and the manner of its consumption have served as a social status and class
marker in society. In India's traditional Hindu culture, the hierarchy of the caste
system "is embodied in what is eaten, with whom, and by whom the meal is
prepared. Food categories and social categories become one: a person acceptable for the table is acceptable also for the marriage bed" (ibid., 156). In medieval Europe, the more expensive and desired animal food remained the preserve of the higher classes (Teuteberg, 1986:17). Not only was the content of the
meal socially determined, but so too was the food service. Social class dictated
precedence in the nobleman's dining hall as well, and a chief usher was typically
there to control who sat with whom, and in what social order guests were served
(Cosman, 1976:105-6). Of course, in modern bourgeois society, status and rank
as a determinant of the quality of food and service have largely given way to the
ability to pay, pure and simple. Cash breaks down all, or almost all, of the old
customary barriers that once limited luxury foods to the nobility and church
hierarchy. Nevertheless, the type of food consumed – fresh salmon, lobster, or fi-
let mignon versus a diet of hamburgers, pizza, or Kraft dinner – and the style of its
consumption – whether in a sumptuous restaurant environ or in a fast-food chain
or family kitchen – are still key markers of social class and economic power.

In recent decades food as a marker of inequality has taken on stark geographical dimensions typically characterized, if somewhat inaccurately, as the
North-South global split. Chronic massive food surpluses in the rich capitalist

countries and chronic food shortages and starvation in substantial areas of the Third World were preoccupying realities throughout much of the 1980s, and they are still present in the 1990s. These realities were and are related in some important ways. In the so-called developed countries, the interests of wheat producers, governments, and large corporate grain traders have coincided around the desire to expand foreign markets. James Wessel's book on the role of U.S. farm exports in the period after the Second World War is a fascinating documentation of this reality (Wessel, 1983). The role of food "aid" to the Third World, such as Public Law 480, which directed U.S. food surpluses abroad, has had objectives and a broad impact far beyond what most citizens of the donor country ever imagined. This "aid" served to tie the host countries to the United States in a number of ways, and it also served to open the door for Third World countries to become buyers of grain from large U.S. grain-trading companies. Indeed, inducements for the conversion of poor nations from recipients to commercial buyers of grain were written into PL-480 contracts from the beginning (ibid., 153). In addition, this food aid generated local currency for the U.S. government, much of it subsequently going to U.S. agribusiness firms in the form of low-interest loans to stimulate their activities in recipient countries. And while food aid was helping to ease political problems posed by chronic grain surpluses at home, and was serving as handmaiden to multinational agribusiness penetration in the developing countries, its impact in the Third World was not as officially intended. The massive importation of commodities such as wheat through the food-aid programs has had an astonishing impact on national diets and basic nutrition. Artificially cheap imported grains compete with the harvests of domestic small farmers, the traditional producers of basic grains. When this is combined with domestic policies that keep prices of local agricultural produce at low levels, the result is the impoverishment of small agriculturalists. In the cities, food "aid" programs have served to restructure the diets of the urban working population as artificially cheap wheat products produced on a mass scale by local agro-industry overcome the variety of indigenous foods that people used to buy from local vendors. By the 1970s and 1980s a diet of soft drinks and artificially "enriched" white bread was supplanting a more diverse range of starches and legumes that had for centuries provided a nutritional balance for the working poor in the developing world. As James Wessel notes, U.S. food exports have undercut local food production in poor countries and created a demand for expensive and nutritionally unnecessary food. Even the World Bank has been forced to admit, "Developed country disposal of agricultural surplus ... tends to accrue largely to a relatively affluent urban minority of consumers, while adverse effects are felt by the poor rural majority" (quoted in Wessel, 1983:176). In considering the food economy, then, we cannot ignore the significant inequalities and contradictions that characterize its consumption, or avoid considering the forces that underlie these contradictions and, indeed, reproduce them in ever more exaggerated and tragic forms.

Food: The Dimension of Production

If the consumption of food has essential physiological, social, and cultural consequences for the human species, what about its production? Here too, it can be argued, food is not a typical commodity. In fact, during the entire existence of modern capitalist societies, which can be dated from the mid-19th century or somewhat earlier according to some interpretations (see Hilton et al., 1978), the production of most of the essential foodstuffs we have consumed has taken place in an "atypical" fashion. What do we mean by this?

At a most basic level, the modern capitalist economy is distinguished from earlier forms of society by the specific way in which material life is organized. Capitalism signified the breakdown of a mode of production centred primarily in small units based in the family household or on artisan labour. This type of production was replaced by a form of enterprise centred around owners and managers, with a labour force severed from its ties to the land and organized around an increasingly elaborate division of labour. Today we consider this form of enterprise to be the "normal" means of providing the goods and services we consume. But oddly, despite the trend in capitalist economies towards large-scale enterprises and away from family or artisan labour working alone or in small groups, the production of food stands apart from the production of most other commodities in that much of it is accounted for by a numerically significant and politically influential population of small commodity producers.[3]

This social class has historically been the source of a populist ideology that has had a notable impact on social development, popular culture, and political process in North America. To some extent it still does today. The contemporary fate of these small commodity producers of food, of the farmers and fishers – the so-called "plight of the family farm," for instance – has special significance for us in a way that is far more profound than our feeling for the decline of independent shoemakers, tailors, or blacksmiths in the face of the superior efficiency of capitalist factory production. Perhaps it is the felt loss of an independent way of life that strikes this responsive chord, but then the same spirit of independence was also associated with most of the traditional craft occupations, and they have not elicited the same kind of social concern. More likely it is the threatened final loss of our historical attachment to the land – a condition that still described the mode of existence of fully one-half of our labour force by the end of the last century (see Manchester, 1983:3). Traditionally this attachment has also entailed a mixing of our labour with the soil in order to produce the plants and animals that allow us to sustain human life itself, and it is perhaps this role as producers of sustenance that helps to maintain a special status, even in this era of battery-hen production and hydroponics, for the family farm and the small-boat fishing population.

The Agro-Food Complex: A Framework for Analysis

If in the past food was not a commodity like any other, distinguished by the social-sexual and cultural significance of its consumption and the non-capitalist form of enterprise that characterized its production, is this still true today? Or has food been reduced to the status of a non-durable consumer good, and its producers – or at least those who still remain – turned into small-business proprietors, pure and simple? These questions lead directly to more all-encompassing issues. Indeed, what has happened to our food economy in recent years? How is it really organized and how did it get that way? Where is it headed? Who have been the main beneficiaries of recent developments, and why? Who are the losers, if any, and what is the nature of the loss?

The objective of this book is to try and shed light on these questions, specifically within the Canadian context, and in doing this it is vital to choose the appropriate framework for analysis. At one point in time it made sense to focus attention on the primary producers themselves, on the characteristics of farming and the organizations producers forged in their attempts to protect and shape the world around them. Indeed, my examination of the early years of the Canadian food economy maintains this focus.

But as we approach the conditions of the present we need a new framework for analysis, because social and economic transformations in recent decades have substantially reduced the importance of the farm population in the wider society, and of farming as an economic activity. For example, by the early 1970s in the United States the net dollar contributions of the farm input and processing industries were some *ten times* that of farming (Buttel, 1980:97). Together with the declining importance of farming in terms of employment and the value-added factor, we now have to account for the massive dependence of farm operators on all manner of agricultural inputs, together with the growing integration of "independent" farming operations and food-processing firms through formal contractual or informal arrangements.

This new reality has stimulated a broader, systemic approach to this sector of the economy.[4] I use the concept of the *agro-food complex* to denote the large number of activities associated with the production, processing, and distribution of food, and with the educational, technical, and ideological apparatuses that provide support and guidance for the more production-oriented activities of the food economy. Within this wider complex it is possible to distinguish various components, or agro-food chains, which taken together make up the whole complex. Distinctive agro-food chains would include the subsector involved with the manufacture and distribution of productive inputs (machinery, chemicals, fertilizers) and the farming activities that constitute the market for firms producing inputs. Another chain involves the production of raw agricultural materials for processing, and their subsequent industrial transformation and distribution. Each of these chains encompasses a specific complex of activities and has its own particular technical and economic features. As one observer remarks, food chains are typically characterized by key control points, representing economic actors with strong bargaining power vis-à-vis other elements in the chain (Scott, 1984:64).

If we are going to truly enhance our understanding of these questions, the inquiry must strive to be both *historical* and *holistic*. It must be historical, because the past inevitably sets the stage for the present and provides a key to understanding the original purpose of current arrangements. By indicating what could have been, the past can also prove to be a guide for what still may be. The inquiry must also be holistic, because the effort to get at the heart of any piece of our social reality will fail if it does not also strive to decipher how the part fits into the whole, and how both part and whole are mutually determined.

The Question of Power

The most significant questions concerning the food economy must also be approached by confronting the question of *power*: who has it, what are they attempting to achieve by it, how are they going about it, and how are social actors, groups, and classes affected by the exercise of the power of other social actors, groups, and classes? With respect to the exercise of power in society, we can assume that any action implies a reaction. Generally speaking, the exercise of power by one group or class, even if it is clearly the dominant group, will produce a reaction by subordinate groups or classes, who act to protect their interests. This "reaction" may, in turn, be reacted upon by the dominant group or class. Ideally, any rigorous attempt to understand social process will be sensitive to the interactive, or dialectical, nature of the process.

The issue of power and of the control it implies, pursued within a framework that is both holistic and historical, provides a central focus for attempting to understand the relationship of the main components of the food economy: the primary producers, the processors, and the distributors of food. The complexity of each of these components must be appreciated. Producers have been increasingly divided by commodity, by region, and by growing differences in the size of operations. Within the processing and retailing industries there are firms of greatly different sizes. There are firms that are privately owned and others that are co-operatively run and ultimately controlled by farmers. There are also crucial sociological distinctions between owners and managers on the one hand, and plant workers and clerical and sales people on the other.

The organization of this book falls into two main parts. The first part deals with the primary producers, farmers: their initial efforts to establish social solidarity, their subsequent struggles to take back control of their economic affairs, and the direction in which these struggles led, almost inevitably, to a nationwide political movement that reshaped our nation. While small-boat fishers constitute the other major group of primary producers, I cannot deal with their struggles here. In the Canadian context, the struggles of people involved in farming and those pursuing fishing have not been closely related, and a separate study would be necessary to do justice to the long and complex history of the fishing population on both coasts.

Part I also considers the attempts of producers to protect themselves from powerful economic actors within the food system and beyond, through legislation that would ensure what has come to be called the "orderly marketing" of agricultural commodities. It also addresses the limitations of the reforms achieved by the agrarian movement in the first decades of the century. These limitations were in part a product of the flawed vision of the farm movement, or at least of the inability of its most progressive wing to subordinate the rest of the agrarian movement to its political project. By failing to achieve reforms in the wider Canadian political economy, farming people ensured their fate as "junior partners" with relatively limited powers in their relationship with an expanding agribusiness complex after the Second World War.

The second part of the book examines the growth and social organization of the complex of activities most commonly referred to as "agribusiness." Surprisingly, it is a sphere that has not been comprehensively studied in the Canadian context, despite its evolution into the largest sector of manufacturing in terms of value of shipments and employment by the 1980s (Canada, 1981:46). Part II considers the forces that have accounted for the food system's substantial shift in power away from the producers and towards capitalist agribusiness corporations. It also explores the power relationships among different subsectors within the agribusiness complex and argues that food-retail corporations have come to play a leading role in the wider Canadian food system.

The final two chapters attempt to provide an understanding of the impact on the agro-food complex of the rapid socio-economic and technological change we are now experiencing, and the political response to this change. In this I pay particular attention to the agro-industrial sphere, and within this sphere I examine the situation of plant workers, who are a very significant, but much neglected, element of our food economy. In their own words, workers in our food business recount how their world has been overturned by the current practice of corporate "restructuring," which for many is a euphemism for plant shut-downs, job loss, long-term financial insecurity, and emotional turmoil. In the final section, as elsewhere in the book, I use a case study approach to provide the reader with concrete details.

In general, I hope the book will introduce readers to a number of the most salient issues in both the development of the Canadian food system and its current realities. My approach is critical, in that it seeks to expose the most serious problems and manifestly irrational forms of social organization embedded in our food system as currently structured. In my view, while some of the problems of our food system are unique to it alone, many of its disturbing tendencies and irrational practices are symptomatic of a wider problem. Through an understanding of our food system, then, we should be able to enhance our understanding of the broader issues that define Canadian reality at the close of the twentieth century.

THE PRODUCERS AND
THEIR STRUGGLE FOR POWER

• • • • • • • •

CHAPTER ONE

BUILDING SOLIDARITY: THE SIGNIFICANCE OF EARLY FARM ORGANIZATIONS

· · · · · · · ·

*In attempting to gain economic security, in fighting for
concrete objectives as solutions to particular problems,
the farmers gradually came to believe that they were
fighting a total system, that the railroads, the Grain
Exchange, the newspapers, all were pitted against them.*

Seymour Martin Lipset, *Agrarian Socialism*

*Class consciousness and group action have been forced
upon us. We accept the inevitable. It is the old battle of the
masses against the classes.*

J.J. Morrison, General Secretary of the United Farmers
of Ontario, *Farmers' Sun*, February 4, 1932

Over the past one hundred years or so, primary producers in Canada have
confronted a variety of other class interests in society: merchants, bankers,
railroad barons, grain handlers and traders, input suppliers, and food-processing and food-retail interests. In these confrontations they have had varying
degrees of success in protecting their own interests and ways of life. Nevertheless, despite the ebb and flow of power, a basic historical pattern has taken
shape – a pattern involving the gradual consolidation of the influence of primary producers through the metamorphosis of farmers' organizations. The

first organization emerged in the 1870s, and the influence of the united producers peaked in the early decades of the twentieth century, followed by a waning of the most visible and dramatic forms of farmers' power. While later in the century producers managed to hold on to some of their gains through the establishment of institutional structures that served their interests, for the most part they have fought a prolonged rearguard action. By the late 20th century the locus of power in the food economy had shifted more and more to the private capitalists controlling agro-industrial and, more recently, food-retailing enterprises.

As a class farmers were once the most numerous social group in Canadian society, and after they became organized they could not have failed to make a profound impact on the development of this nation. A profound influence they did have, although the end results of their struggles did not always meet their original expectations.

The Grange: The First Attempt to Consolidate Producers' Power

The relative power that primary producers enjoyed at different times can only be understood as the result of their dynamic relationship with other classes, a relationship that was itself shaped by the different historical phases of the development of industrial capitalism. In the late 19th century, this contingency was most evident in the United States during a crucial period in the expansion of U.S. capitalism that came with the building of railroads. It was not a coincidence that the first mass-based farm organization in the United States – the National Grange of the Patrons of Husbandry – took root at a time when rapid concentration and trust formation were being orchestrated by U.S. railroad capitalists.[1] The manipulation of freight rates to the disadvantage of Western farmers dependent upon Eastern markets pushed mid-Western agriculturalists to make their first notable forays into the political arena. Through the agitation of "The Grange," numerous states were forced to pass legislation curbing the worst excesses of the railroad barons (see Wood, 1975:25-26).

In the United States the Grange also offered an alternative to what farmers viewed as excessive profits captured by middlemen who marketed their crops in Eastern markets. This alternative was in the form of early co-operative organizations, which endeavoured to purchase inputs directly from manufacturers and find more extensive markets and therefore better prices for their crops (ibid., 26).

In Canada the early Grange organizations were instigated by Eben Thompson, an American. Initial success in the Eastern Townships of Quebec was followed by an even greater receptivity to the Grange in Ontario. Before long Canadian farmers were concerned with the need to establish a sovereign national organization, with independence from the U.S. mother organization. In 1874 the Dominion Grange was given birth in London, Ontario (ibid., 41-43). Membership peaked at about 30,000 a few short years later, and thereafter the Dominion Grange went into decline.

Canadian farmers combining in the Grange organization did not achieve the dramatic results experienced by their U.S. counterparts. In Canada the Dominion Grange encouraged the development of only a mild and narrowly based agrarian populist ideology, which limited the scope of political action. As Louis Aubrey Wood notes, in its declaration of principles the Dominion Grange showed no particular hostility to big business *per se*, even as it affected the farming community: "On the other hand, the Grange in Canada is firmly determined to bring producer and consumer more closely together. 'We must dispense with a surplus of middle-men,' contends the document, 'not that we are unfriendly to them, but we do not need them – their exactions diminish our profits.' Any 'tyranny of monopolies' would be distasteful to the granger just as he would be 'opposed to excessive salaries, high rates of interest, and exorbitant percent profits in trade'" (Wood, 1975:45-46).

This hostility to middlemen and to the excesses of business in the form of trusts and monopolies was tempered by a definite reluctance to enter the political arena. The Dominion Grange's founding document stated, "No grange, if true to its obligations, can discuss partisan or sectarian questions." This aversion to any direct political activity, together with a restriction on membership to agriculturalists alone, undoubtedly restrained the potential of the Grange to serve as an effective vehicle for the advancement of the interests of primary producers. That vehicle would only move forward with the development of new organizational initiatives elsewhere in Canada.

Wheat Producers Organize

The rapid rise and fall of the Dominion Grange were but a prologue – though a necessary period of experimentation – that soon gave way to more concrete expressions of the power of primary producers. By the turn of the century this power was taking several concrete forms: pro-farmer federal legislation; a rapidly developing, producer-controlled co-operative movement; and, somewhat later, political parties that explicitly represented farm interests.

Canadian farmers first became a tangible social force in society in the Prairie region. For several important reasons, their influence also took on its most radical and enduring forms there. The initial struggles were provoked by the emergence of several large grain-elevator companies in the 1890s and the decision of the Canadian Pacific Railway to provide these elevator companies with exclusive loading rights in all their established locations. This decision threatened to eliminate the competition provided by the many "flat" warehouses owned by independent grain traders throughout the Prairies – warehouses that had previously provided farmers with several buyers for their wheat at each collection point. As evidence of collusion by the elevator companies mounted, wheat farmers and independent warehouse owners alike organized a vigorous protest.[2]

One thing is fairly clear about these early protests: the federal government was disposed to treat them seriously and to move to diffuse the situation with surprising speed. Very soon after the outcry over the grain-elevator situation, the federal government appointed a royal commission to examine the issue and seek solutions. The findings of the commission confirmed the charges of the agrarian interests, and the text of its report is a classical statement of the types of restrictive practices that were welding together the farm community in these times.

> As a result of the refusal of the railway companies to take grain from a flat warehouse (which resulted in driving many small buyers out of the market) and of their refusal until 1898, to furnish cars to farmers desirous of doing their own shipping ... the elevator owners have had it in their power to depress prices below what in our opinion farmers should realize for their grain. It would naturally be to their interest to so depress prices; and when buying, to dock as much as possible....
>
> There being no rules laid down for the regulation of the grain trade other than those made by the railway companies and the elevator owners, we think it of great importance that the laws should be enacted and that rules should be made under power given by such laws, which will properly regulate the trade.[3]

While we are now more disposed to view royal commissions as a mechanism by which the party in power diffuses an issue without ever having to really deal with it, the experience of these early agrarian commissions was surprisingly different. As Vernon Fowke (1948:169) argues, "It would be difficult to find commissions anywhere whose recommendations were more promptly or more completely enacted than those of the farmer-dominated royal grain inquiry commissions of 1899 and 1906." This prompt enactment of commission recommendations responding to agrarian protests came, not coincidentally one might surmise, on the eve of two separate federal elections; and it was not an isolated response of the federal government. Whenever a crisis brought forward some form of serious protest from farmers, whether it was simply pressure on local Members of Parliament or mass marches on the House of Commons – like the march in 1910 with representation from farmers from across the country (but notably the West and Ontario) – legislation to address the grievances was not long in coming.

How do we understand the intensity of farmer protests at the time and account for their success in bringing forward legislation to redress their concerns? For one thing, the business class at the turn of the century operated in a relatively laissez-faire environment, unencumbered by a government regulatory apparatus that would curb the worst excesses of private capital in the use of raw economic power. Turn-of-the-century capitalism was decidedly nasty in many respects, and in their thrust to control markets powerful entrepreneurs dealt with the interests of other sectors of society, and notably the large farm population, in what we would consider today a highly cavalier fashion.

Within the farm community some producers were much more vulnerable than others. This was especially true of wheat growers in the West. As Seymour Lipset (1968:48) and others argue, wheat farmers were almost totally dependent on a crop that was prone to extremely volatile price fluctuations.[4] This dependency had much to do with the more radical protests that emerged in the West in response to the railway, banking, and grain-trading interests that were based largely in the East. In addition, the fact that they tended to produce a single commodity brought a relatively high degree of unity to their cause, despite their disparities in wealth. Finally, there is evidence to suggest that before their emigration the majority of the newly settled prairie farmers had shared similar class experiences in Europe as skilled and unskilled labourers. Many had undoubtedly been exposed to radical labour movements in the old country. Certainly, those who emerged as the leaders of farmers' movements in the West had typically been involved in socialist politics either in Britain or the United States (Lipset, 1968:43).

The political circumstances of the time forced the two main federal parties to be sensitive to the complaints of the farm population, and especially to those of the relatively unified Western agrarian interests, who carried an influence in determining political fortunes in Ottawa.

Discovering the Power of Organization

While on paper farmer agitation achieved substantial gains – such as the Manitoba Grain Act – in practice the provisions of legislation were often flagrantly violated. Wheat farmers found that while they now had laws guaranteeing the rights of the small grain warehouses to do business alongside the big concentrated elevator companies, it mattered little because at harvest time the CPR steered the bulk of its grain cars to the elevator companies anyway (see MacIntosh, 1924:13). In his classic book on farm movements in Canada Wood states, "Grain growers averred that the existence of a loading platform at any shipping centre should tend to stabilize prices at that point, but as the elevators were receiving most of the cars they were in virtual control of the market. From data obtainable it would appear that the forcing down of prices was at its worst in those localities where no independent buyers came into competition with the elevators" (Wood, 1975:178).

With wheat prices showing extreme volatility, and with grain traders, elevator companies, and railroads playing fast and loose with the fortunes of farm operators, the feeling of Western producers was that they were being forced to the wall. A Saskatchewan farm leader of those times, W.R. Motherwell, described the temper of the times: "There was incipient rebellion when we organized. It's too late for organization; it's bullets we want, men were saying. But we really didn't know what we wanted; we were in despair. It was not a question of growing crops but of marketing them. In the fall when the elevator opened, you would see a rush of wagons, wheel to wheel, to see who would get there first.... Such conditions engendered bitterness and the country was ready for anything" (quoted in Lipset, 1968:59).[5]

Farm leaders in the West were particularly perceptive in their analysis of what was needed to meet the challenge coming from those whom they considered their economic foes. As an early observer noted, they understood that "organizations such as the railway and the NorthWest Elevator Association could only be met with counter organization" (MacIntosh, 1924:14). This belief, together with the real anger and frustration of many farmers, gave impetus to the formation of the Territorial Grain Growers' Association in 1901, later to be called the Saskatchewan Grain Growers' Association. Similar farm organizations had already been set up in Manitoba and Alberta.

The early work of these organizations was to see to the enforcement of the existing legislation that farmers had agitated for and, where the legislation was deficient, to bring pressure to improve it. Later on these goals would change. The associations of grain growers were remarkable for the impetus they provided in the creation of perhaps the most powerful and enduring achievements of primary producers in the Canadian context: the co-operative movement and producer-controlled co-operative enterprise.

Fighting Combination with Combination

Competition is the law of destruction, and all the destruction that has ever been wrought by man against his fellow men has been wrought by competition.... All construction of social strength has been done by co-operation.... As soon as we begin to develop co-operation for co-operation's sake ... then will begin the real test of the ability of the human race to become truly civilized.

Henry Wise Wood, United Farmers of Alberta

Early initiatives in forming producer-controlled enterprises were notably linked with the name of E.A. Partridge, a farmer-leader Louis Aubrey Wood describes as "a big man, of restless demeanor and flashing eye." According to Wood, Partridge "became the seer of the Canadian plains-folk. More ideas have originated with him affecting the farmers' social and economic welfare than with any other dweller in the grain country" (Wood, 1975:183). In 1905, supported by farmers in the area of Sintaluta, Partridge ventured to Winnipeg to conduct a month-long investigation of the Winnipeg Grain Exchange, and he came away with a greater appreciation of the wheat business, which he described as being "practically in the hands of three milling companies and five exporting firms" (quoted in MacIntosh, 1924:19). This was much to the detriment, he believed, of the farmers' interests. His views on the workings of the grain trade were supported some years later by the report of the Saskatchewan government's "Elevator Commission" of 1910.[6]

Partridge also left Winnipeg convinced that farmers could combine themselves as marketers of grain and thereby distribute among themselves the profits that were instead being captured by a small group of grain merchants. In a letter he circulated in 1905 promoting his ideas, he wrote, "A thousand farmers controlling ten million bushels of wheat and selling through a single accredited agent would be in a position of a single person owning ten million bushels. It is a well known fact that the owner of ten thousand bushels can make a much better bargain for his wheat than the owner of one thousand bushels. How much would this power be augmented in the owner of ten million bushels?"[7] His ideas about forming a co-operative farmer-owned company for the handling of grain were not universally embraced by grain growers – not surprisingly, given that the early efforts of farmers to form a co-operative enterprise associated with the Dominion Grange were widely viewed as disasters. But there was enthusiasm among the farmers of the

E.A. Partridge
1861–1931
Farmer, teacher, agrarian activist, and visionary

Edward Alexander Partridge was one of the most significant personalities in shaping the history of the Canadian West in the early decades of this century.

Born in the village of Dalston, Ontario, Partridge taught school in that province before migrating to the prairies with his brother in 1883 to take up homesteading near the town of Sintaluta, Saskatchewan. There he farmed, taught school, and raised a family.

Despite experiencing great tragedy – he lost his daughter and two sons through accident and the First World War, and lost his leg in a farm accident – Partridge continued to stand at the forefront of the emergent agrarian movement in the West for much of his life. He was a founding member and director of the Territorial Grain Growers Association (1901), a major first step in the establishment of effective agrarian collective action. His vision of co-operative associations to replace the tightly controlled, privately owned grain trade was particularly influential. Partridge's study of the Winnipeg Grain Exchange in 1904 led to a concrete proposal to set up an alternative producer-owned and controlled co-operative grain company.

Sintaluta region, and it was there that the Grain Growers' Grain Company, with its motto "In Union Is Strength," took form (Wood, 1975:185).

Although farmer-controlled co-operative enterprises subsequently came to hold considerable economic clout in their own right, their establishment was originally viewed as a stepping-stone to government ownership. The principle that government should purchase existing grain terminals and establish a network of storage elevators was held by the Grain Growers' associations in both Manitoba and Saskatchewan (MacIntosh, 1924:34). In fact, in their annual conventions the Grain Growers' associations had also passed resolutions calling for the nationalization of the telephone system and provincial coal and oil resources. Later they added public utilities, transportation, and banking to the list (Lipset, 1968:73).

The history of government involvement in this area leaves little doubt

Despite widespread scepticism and outright opposition, Partridge's singular determination won out, and his idea to fight the combination of the private railway monopoly and grain trusts through producer-controlled co-operative enterprise was realized in the establishment of the Grain Growers' Grain Company. Before long, co-operative grain enterprises controlled much of the Prairie grain industry. Later on he took the lead in pressing for a government-owned public elevator system.

A man who personified the advance guard of agrarian thinking, he was acutely aware of the need for education and communication to develop farmers' understanding of their material circumstances and an awareness of possible long-term solutions to their problems. With this objective in mind he organized and was first editor of the *Grain Growers' Guide*, which came to be a crucible for nurturing progressive agrarian thinking and fostering solidarity among the grain growers.

The operation of the Canadian political and economic system, the practices of the private grain trade, and the inability of the class of numerous primary producers to influence the two entrenched old-line parties eventually pushed Partridge to develop a more radical and profound critique of the whole system. His attacks on the rapacious capitalism of his time and his call for a new social order built upon the principals of human co-operation helped inspire efforts to build an alternative agrarian-labour party – the CCF – during the 1930s. Partridge was a visionary, but one with the capacity to transform far-reaching ideas into practical, workable enterprises that served the grain growers' interests.

Source: Knuttila (1989); Lipset (1968); and Laycock (1990).

about the influence of these associations of primary producers on the political life of their respective provinces. Despite considerable provincial government resistance to involvement of the state in any way that would prejudice the big grain-handling and grain-trading companies, the concerted pressure of the grain-growers' organization – given force by a massive petition signed by 10,000 farmers (Wood, 1975:210) – turned around Manitoba's government led by Conservative Rodmon Palin Roblin. In 1909 it moved to "accept the principle laid down by the Grain Growers' Association of establishing a line of internal grain elevators as a public utility" (quoted in ibid., 37).

As Lipset argues (1968:67), the Roblin government was not so much interested in taking action on this plan as in conceding superficial concessions with an eye on the next election. Indeed, given a series of ill-conceived decisions in the plan's implementation, Manitoba's experiment in government ownership of a substantial part of the provincial economy did not fare well.[8] Perhaps its greatest significance was to turn efforts elsewhere, especially in Saskatchewan, towards producer-controlled co-operative organization as a favoured way of controlling the grain-handling system. In 1911 the Saskatchewan government passed a bill to create the Saskatchewan Co-operative Elevator Company, which would store, buy, and sell grain (Wood, 1975:213). Under the new co-operative company, stock could only be sold to farmers, and provisions were introduced to limit the number of shares owned by any one individual to ten (Wood, 1975:213).

With the birth of producer co-operatives in the grain business, Canadian farmers, or more specifically Western farmers, made a historically important advancement in controlling their own destiny. And by taking on the business of organizing the domestic and overseas sale of wheat, these co-operatives moved into a position to compete head to head with the large privately owned grain-elevator companies (Wilson, 1978:55). Between 1907 and 1917 the percentage of grain handled by the farmer-controlled co-operatives rose at an impressive rate, from approximately 3 per cent to about 35 per cent of all grain handled on the Prairies (see MacIntosh, 1924:Appendix B). While making these major inroads in the grain-handling business, the Alberta and Manitoba co-operatives were also moving into the co-operative purchase of farm supplies and machinery, into publishing, and even into the establishment of a co-operative sawmill (Wilson, 1978:55). Before long, Canadian grain growers had organized the largest co-operative elevator system in the world, a system that also bought and sold grain (Lipset, 1968:69).

The 1920s saw a further impetus given to this co-operative system – an impetus once again provided by the failure to establish a state-controlled enterprise. This time it was the federal government's reluctance to set up a permanent marketing board for wheat as an alternative to the futures-marketing system run by the speculators of the Winnipeg Grain Exchange.[9] In the absence of state intervention to provide a central selling agency to avoid speculation in wheat as a commodity, and given the volatile fluctuations in price that this inevitably produced, the leaders of the Prairie farm organizations quickly turned to a pooling system for wheat.[10] A pooling system meant that the farmers who participated were able to avoid the Grain Exchange because they had their own central agency to sell wheat directly through offices established in Europe, South America, and even China (Wilson, 1978:220-21). It also made sense to bring together the farmer-controlled elevator system with the selling function of the "pools." By the late 1920s three provincial farmer-owned wheat pools controlled about one-half of the total grain storage capacity in Western Canada. This represented an impressive advance in barely two decades from the old situation, in which a "combine" of a few private elevator companies, together with the complicity of the CPR, had determined the fate of tens of thousands of Prairie grain producers.

From Co-ops to Political Intervention

The earliest forms of producer organizations, such as the Dominion Grange, were at least in part an attempt by farmers to gain more control over their material circumstances. The associations of grain growers in the West took this matter quite seriously, pushed as they were by the elevator combine and the CPR and their own vulnerability as producers dependent upon the ups and downs of one commodity, wheat.

Farmers found that "getting organized" certainly helped bring the attention of the powerholders in society to their concerns, but it usually did not prove to be enough to protect their interests in any concrete and significant way. Western farmers seemed to have the clearest vision about how to change this situation. Their leadership understood that the interests of the broad mass of producers could only be served via changes in the institutional structures of society. In their case, the key institutions – the banks, railroads, grain-storage companies – were all privately owned and, by the turn of the century, controlled by a very small group. Everyday experience taught farmers that their interests, and the interests of those who controlled the organizations they were forced to deal with, were not necessarily in concert. Indeed, the interests were often diametrically opposed.

The most forward-thinking farm organizations looked to the state, which they understood as the expression of the "public interest," to provide an alternative to the privately owned financial, rail, and grain-handling "combines." A frustration with their initial experience with state ownership in Manitoba pushed grain producers on the Prairies in another direction: farmer-controlled co-operative enterprise. Co-operative development in the West was thus a historically significant event for shifting the locus of control into the hands of the most populous sector within the food economy: the primary producers.

Co-operatives helped primary producers to avoid some of the most rapacious aspects of this early period of Canadian capitalism, and with arrangements such as the pooling of wheat, they helped ensure that farmers were dealt with on an equitable basis by the organizations in charge of handling and selling their wheat. They wished to avoid the prevailing situation where there was a deferential treatment of farmers according to their location or the size of their operation.

A degree of control and equity: this was a positive gain, but co-operatives did not solve all the problems faced by producers. In the West in particular there was a growing belief – especially as the economy degenerated into what came to be known as the Great Depression – that the magnitude of the problems required an active intervention into the political realm of society. While this belief would have its most enduring and concrete manifestation on the Prairies, the precedent of an agrarian political party in power would be set further away, in Ontario.

CHAPTER TWO

THE ROLE OF FARMERS' PARTIES: A NEW FORM OF POWER

• • • • • • • •

We suffer not from visionaries, but from those who lack vision. It cannot be too often repeated, "Where there is no vision, the people perish."

E.A. Partridge

In one sense, Ontario was an unlikely place to witness the debut of a government ruled by an agrarian party. Only a few years before the 1919 victory of the United Farmers of Ontario, producer organizations had been a widely scattered lot. W.C. Good, one of the foremost agrarian leaders in the province at that time, later noted that Ontario farmers had been organized diffusely into "a very considerable number of locals of the Dominion Grange and Farmers' Association; a considerable number of Government sponsored Clubs; and quite a number of special purpose organizations having to do with live stock, fruit growers, seed grain, etc." (Good, 1958:93). Although he added, "In a few places there was at least the nucleus of local general purpose cooperatives," the situation was a far cry from the relatively centrally organized Prairie grain-growers' associations.

Inspired by the grain-growers' movements in the West and keenly aware that more than one farmers' organization in Ontario had recently withered on the vine, the most progressive farm leaders of the province set out to build a truly unified and broadly based farm organization in 1913 (*The Farmer's Magazine*, September 1952). The following year, as World War I began, the United

Farmers of Ontario (UFO) had its founding convention in Toronto. Led by such notables as J.J. Morrison, Ernest C. Drury, and W.C. Good, the United Farmers was to provide the organizational vehicle for the agrarian protest that welled up in the province several years later.

The Farmer-Labour Government in Ontario

As the First World War brought on calamity for millions of people, it also played a major part in propelling social movements forward. The most momentous movement was perhaps in Russia, where the damage done to the social fabric of the Czarist state proved irreparable and an upsurge of rural unrest helped bring down the entire edifice in the Bolshevik-led revolution of 1917. In Canada, somewhat less dramatically, the war also brought an upsurge in rural protest that ended in farmers coalescing to form the government of the most economically significant provincial jurisdiction in the country.

The issue that brought rural resentment to a boil was conscription. In 1917 the federal Conservative government brought in the Military Services Act to conscript males between the ages of twenty and forty-five. Farm organizations, which had generally opposed this idea, were not long in reacting. A proclamation coming from the UFO's third annual convention set forth the organization's position in no uncertain fashion: "Since human life is more valuable than gold, this convention most solemnly protests against any proposal looking to the conscription of men for battle, while leaving wealth exempt from the same measure of enforced service. It is a manifest and glaring injustice that Canadian mothers should be compelled to surrender boys around whom their dearest hopes in life are centred, while plutocrats, fattening on special privileges and war business, are left in undisturbed possession of their riches" (quoted in Wood, 1975:277).

Despite these defiant words, farm communities were lulled into a state of passive acceptance of the legislation when they were told their sons would receive exemptions from military service because of the vital role the boys were playing in producing much needed foodstuffs for the war effort. This promise did not hold. Pressure on the Canadian government to bolster its commitment of troops for the war effort in Europe persuaded the government of Robert Borden to end the policy of exemptions, just as farmers across the country were putting another year's crop into the ground. Farmers from Ontario and Quebec rapidly organized and sent 5,000 of their number to Ottawa to put their case before the government. Although they met the Prime Minister, he refused to budge on the matter of exemptions, and he later blocked their representations to the House of Commons. This rebuff sent shock waves through the rural communities of Ontario and Quebec.

In Ontario a special convention of the UFO was organized a short time later to consider the situation, and out of this large and tumultuous meeting there emerged a call for independent political action. Soon after the UFO contested and won a by-election on Manitoulin Island. Moreover, the membership of the UFO began to soar, moving from 2,000 members in 1915 to 48,000 in 1919 and reaching 60,000 a year later (Wood, 1975:ch.3).

In the provincial election of 1919 the UFO won forty-four seats while the incumbent Conservative government, tainted by its association with the unpopular federal Conservative government, barely managed twenty-six. The unexpected success of the farmers' candidates brought the UFO to power before it had even decided on a leader (Johnston, 1988:629).[1] As the largest group in the legislature, but lacking a majority, the farmers' party needed to forge a coalition with another political party to form the government. This it did with the small Independent Labour Party, and after the UFO secretary J.J. Morrison declined the invitation, it selected a noted agrarian voice, Ernest C. Drury, as leader.

The farmer-labour government was short-lived, but it did manage to advance a reform agenda, its record a testimony to the potential inherent in an organized agrarian voice. The more prominent achievements included:

- recovering for the public treasury considerable sums of money as compensation for the fraudulent allocation of timber licences by the former Conservative government.
- increasing the stumpage dues paid by the forest industry exploiting Crown lands by 100 per cent.
- doubling the amount of money collected from the very rich in the form of succession duties.
- increasing the maximum percentage payable to injured workers from 55 per cent to 66.6 per cent of their average wage, while also increasing the allowances to widows, invalid husbands, and their children.
- enacting a minimum-wage law for women and girls.
- making funds available for public-health nurses and free clinics throughout the province (*Farmers' Sun*, June 9, 1923).

The government also engaged in a long struggle to bring the Hydro-Electric Power Commission, the forerunner of Ontario Hydro, under tighter fiscal control, because the Commission was threatening the state treasury with massive cost overruns. Much of the government's legislation had a broad appeal, but it attempted to serve its rural social base with an impressive expansion of the county road system in Southern Ontario as well as a network of new roads in Northern Ontario.

In addition the government established a provincial savings institution to generate revenues to provide a provincial farm credit system to aid with improvements on Ontario farms. This legislation was vigorously opposed by the Canadian Bankers Association, which organized a campaign to protect privileges the banks had previously regarded as theirs. The Minister of Agriculture, Manning Doherty, launched a rapid counterattack in which he accused the Bankers Association of deliberately attempting to mislead the public. "Our banker friends," he noted, "have so long in this country enjoyed special privileges that they have now come to look upon these, not as privileges enjoyed at the will of the people, but as rights of the banking fraternity" (*Farmers' Sun*, March 23, 1923).

The coalition government was also able to move ahead with legislation that impinged on the prerogatives of the big manufacturing and banking interests. Nevertheless, there were established limits on how far its members were able to push their cause for a more equitable and fairer reorganization of the provincial economy. Legislation that threatened to do something about the "combines" that lessened competition was a case in point. The large retailers and manufacturers launched a campaign that effectively stalled legislation developed to deal with price fixing: they argued that the government had no right to interfere with the "immutable laws of business" (*Farmers' Sun*, October 19, 1922).

Alongside its achievements and partial victories, the government succeeded in welding together an alliance of farmer and labour forces to advance common goals. This alliance was not without its problems, and it began to disintegrate towards the end.[2] It did, nevertheless, establish a precedent of some magnitude for the political future of the agrarian movement and its relationship with the organized section of the Canadian working class.

The reform process led by the agrarian forces might have advanced further had it not floundered on the shoals created by internal dissension and a hostile, Toronto-based press. The internal struggle between factions led by government leader E.C. Drury on the one side and the eloquent farm activist and UFO secretary J.J. Morrison on the other cast a long shadow over the government's last year in power. The debate centred around Drury's proposed "broadening out" strategy, and the polemics it engendered indicate the difficulties faced by agrarian interests in charting an independent and effective course in the political life of Canadian society.[3]

By arguing for a "broadening out" strategy, Drury sought to extend the social base of the agrarian movement and thereby avoid having the movement
branded as representing only one class in society. He was fundamentally opposed to a strategy based on class politics and advocated a political orientation
that would accommodate various class interests around certain political principles or objectives. His office stirred up considerable dissension when it sent
out a "confidential" letter, mostly to urban constituents, soliciting names of
persons who might be willing to form a new Progressive Party in the province
– a party that would presumably attempt to capture some of the urban vote as
well as the farm vote already supporting the UFO. After it was published in the
Toronto *Globe* on August 1, 1922, the letter received a hostile reaction from
leaders of the United Farmers of Ontario, who saw the proposed party as a
threat to the farm movement as then constituted.

Led by UFO Secretary J.J. Morrison, but also supported by the widely dis-
tributed newspaper of the movement, the *Farmers' Sun*, this internal opposition adamantly proclaimed that Drury's views were likely to transform the
agrarian movement into a political vehicle that would be all too similar to the
old-line Tory and Liberal parties.[4] In their view the old-line parties had historically repressed a politics based upon real class interests in society and
brought to the fore issues such as ethnicity or religion that polarized voters
along non-class lines. Morrison and the others argued that by diffusing and
suppressing a politics based on class issues, the agrarian movement would lose
its hold on the interests of the productive classes in society – principally farmers and the urban working class in their view – with these interests ultimately
subverted to the designs of the powerful few who controlled Canadian society. They felt that a primary purpose of the movement was precisely to raise
the class consciousness of agrarian producers.[5] Rather than abandon its agrarian roots, the UFO should stick to the principles that farmers had hammered
out in democratic convention, they argued, while welcoming and encouraging alliances with classes and social groups that were sympathetic to and could
support those principles. In his strong attack on the Premier's controversial
letter, Morrison proclaimed that the UFO "was a spontaneous protest of the
shortcomings of the old party system that came from the rural people, and because it came from the farmers themselves, the farmers, and farmers only,
have the right to direct its future (*The Globe*, August 16, 1922).

Not long afterwards, UFO members in convention supported Morrison's stand
against a broadening out policy. But in the months that followed, press coverage
of the continuing split between Drury and UFO secretary Morrison plus the de-

James J. Morrison
1861–1936
General Secretary, United Farmers of Ontario

James Morrison, known to the public as J.J. Morrison, was a man *The Toronto Daily Star* once described as "the kingmaker and wrecker of governments," a man who "for nearly four years has sought to establish himself as dictator of Ontario" (April 18, 1923). Morrison was the son of Robert Morrison, an Irishman who had despaired of the condition of his rural countrymen under the rule of English absentee landlords and had travelled to Canada at age twenty-two to set himself up as a pioneer farmer in Wellington County, Ontario. His mother was the daughter of a woman with reform leanings and sympathies for the Mackenzie rebellion.

"J.J." left the farmstead at age twenty-five to seek a business training in Toronto, where he worked for fifteen years at a variety of employments before returning to take over his father's farm. Morrison was a catalyst in 1913 in establishing the United Farmers of Ontario and thereby securing for the rural producers an effective organization that could advance their interests in a way the Grange had not been able to. Soon after, Morrison was elected Secretary of the UFO. It was through this organization and through the pages of its newspaper, the *Farmers'*

fection of some UFO MPPs to the Liberals (see *The Globe*, January 9, 1923) prevented a closing of the wound. As an election campaign rolled around, the UFO's enemies made the division in its ranks into a public issue, and there were rumours that the Premier was contemplating a fusion with the Liberal Party. These factors no doubt played their part, and overtures to townsfolk by Premier Drury's government, through legislation and otherwise, seemed to carry little weight. The farmer-labour government went down to defeat in June 1923, never to reappear. In the end, farmers in Ontario had not been convinced that the government was theirs and that it deserved their sustained loyalty.

Agrarian Politics after 1923: The National Progressive Party
The quick demise of the farmer-labour government in Ontario is not altogether surprising, especially given the level of organizational and political

Sun, that Morrison's influence spread with the rising tide of agrarian protest after World War I.

Described by the national magazine *Saturday Night* as "one of the keenest political minds the country has ever produced," Morrison was offered the premiership of Canada's most powerful province after the UFO victory of 1919. He turned it down. He is also credited with blocking the bid by the powerful Sir Adam Beck, a Conservative and the force behind the creation of what became Ontario Hydro, to accede to the office of premier at the head of the UFO government. During the years of the farmer-labour government Morrison challenged Premier Ernest Drury's attempt to "broaden out" the UFO and make alliances with the Liberals. Morrison rallied the farmers at the annual UFO convention to defeat Drury's plan, but the confrontation left deep scars that were to prove costly at the next election.

Together with Henry Wise Wood in Alberta, J.J. Morrison felt strongly that the practices of the party system were fatally blocking the emergence of a truly democratic political process, one that would allow the farm community a voice in keeping with their numbers. Both were advocates of "group government." Morrison was tireless in his efforts to increase the class consciousness of farmers and thereby enhance their ability to advance their interests in society. In the words of John Kenneth Galbraith, Morrison "was one of those who made the revolution in farmer attitudes which will always be the mark of the first decades of this century."

Source: B. Bystander, "A Pen Picture of J.J. Morrison, the Farm Leader of Ontario, and General Secretary of the UFO," *Saturday Night*, April 21, 1923; *Toronto Star*, April 18, 1923; "Scrapbook of J. Morrison," Archives of Ontario, file no. MV 7125; Galbraith (1956).

maturity of the agrarian movement in Ontario as compared to that elsewhere, and especially the Prairies. Given the UFO's rapid growth in the few years before 1923, it had little time to put down the deep roots that would be needed in the political terrain. Furthermore, the always thorny question of leadership and the defection of agrarian party leaders to the ranks of the Liberal Party were constant obstacles to solidification of an agrarian party in the West as well as in Ontario. The difficulties facing any attempt to create a political vehicle with an authentic agrarian focus became most evident at the federal level in the 1920s with the appearance of the National Progressive Party and the struggle over the perennial agrarian bugbear, the national tariff policy.

In the 1920s farm organizations were pressing to achieve a downward shift in the duties for goods that they depended on for production: items such as farm implements and machinery, fertilizers, and lubricating oils. For decades they had paid a premium on these goods, ostensibly to give protection to Canadian manufacturers against U.S.-based competition. In response many U.S. companies had simply "jumped" the tariff wall by setting up branch-plant operations in Canada and crowding out local capital through their on-site operations. Farmers believed that the real effect of the tariff wall was to reinforce the tendency to combination – or corporate concentration, as we would call it today – and there is good evidence that this concentration did take place, due to the tariff and/or other factors (see Figure 1).

Before 1920, to the extent that the numerous Canadian farm organizations had a unified view, it was to be found in the Canadian Council of Agriculture. The Council served to focus the issues that the disparate farm organizations had in common. By 1917 the Council had helped forge a "Farmers' Platform" (see Morton, 1950: Appendix B). Much of this agrarian program centred on the need, from the farmers' point of view, to reform Canadian tariff legislation, which had originally evolved along with the National Policy of John A. Macdonald's Conservative Party decades earlier.[6] However, the Council was typically cautious about taking direct political action to further these ends, favouring a concerted movement to pressure the old parties to adopt pro-agrarian policies instead. This position was not surprising given the influence of the prominent grain-growers' associations on the Council and the close ties of their respective leadership with the Liberal party in many instances.[7]

The grassroots of the agrarian movement, and its more radical leadership, favoured independent political action. This option blossomed with a melting away of the nationalistic fervour associated with the First World War and the resurgence of deep-seated problems plaguing farm communities in Canada, notably rural depopulation, drought in the West, and declining wheat prices (Morton, 1950:66). Another potent factor was the years of patient agitation of the more radical wing of the farm movement, utilizing such widely read agrarian organs as the *Grain Growers' Guide*. In addition, as Morton has forcefully argued in a classic study of this period, the old-line parties were decisively losing their hold over rural folk, especially in the West. The Conservatives' prospects had been badly damaged by their defeat of the proposed Reciprocity Agreement with the United States in 1911 and their continued association with protectionist policies that were widely disliked by the direct producers. Liberal prospects had been hurt over the issue of conscription in 1917. The party had been badly split between anti-conscriptionists centred in Quebec and other members in favour of conscription and a wartime coalition with the Conservatives.[8] Finally, the rise of a national agrarian party was propelled by

Figure 1
Industrial Concentration in the 1930s by Control of Output in Selected Manufacturing Industries

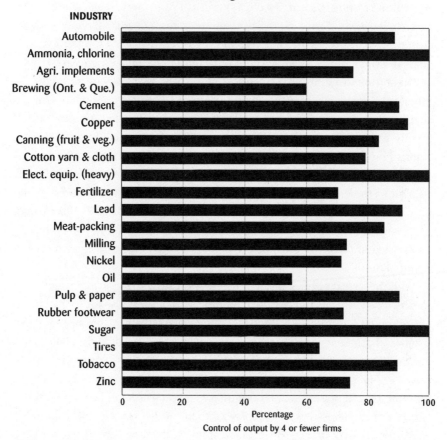

INDUSTRY

Percentage

Control of output by 4 or fewer firms

Source: Lloyd O. Reynolds, *The Control of Competition in Canada,* 1940, Table 1:5; in Clement, 1975:84.

one further factor: the impact of the UFO victory in Ontario in 1919. While agrarian protest had earlier had a largely geographical or "sectional" character, the UFO success demonstrated the more broadly based, national nature of the agrarian movement and reinforced the idea that it represented the aspirations of a powerful class in society, not just a regionally based protest movement.

After World War I the independent agrarian members of the federal Parliament were expecting some relief on the tariff by the Borden government. When little was forthcoming in the first postwar budget, they decided to stake out a position more clearly in opposition to the ruling Conservatives. Previously they had participated in the wartime administration of Borden, and their nominal leader, Thomas Alexander Crerar, had served as federal Minister of Agriculture during the war. By-elections in 1919 saw some major gains for the agrarian interests. That year the convention of the Canadian Council of Agriculture gave a strong vote to revive the fight to bring down the tariff wall. A year later the Council facilitated the birth of the National Progressive Party.

Right from its founding convention, the Progressive Party mirrored the internal tensions that were to pry apart the farmer-labour government in Ontario shortly afterwards. While farmer delegates from Ontario and Alberta wanted a truly independent political presence in Ottawa, explicitly representing agrarian interests, others, including Crerar, favoured a party that was not rigidly agrarian in its stance (Wood, 1975:352). In the months afterwards Crerar took pains to deny that the party was formed to serve a class purpose, while provincial agrarian leaders like J.J. Morrison in Ontario and Henry Wise Wood, leader of the United Farmers of Alberta, were talking another story, one in which the reinforcing of the farmers' consciousness of their class interests was a central motif.

With the defeat of the Conservatives in 1921 the Progressive fortunes blossomed. The party elected sixty-five members from across Canada to the House of Commons. Of those MPs, fifty-two were active farmers (Morton, 1950:148). Their strength was surprising because the victorious Liberals had elaborated an electoral campaign that borrowed much from the farmers' platform, notably promising substantial tariff reform (Wood, 1975:356). However, after refusing to form the official opposition and opting for a special accommodation with the Liberals, the Progressives were to play a difficult and ultimately self-destructive game (for them) of holding the balance of power in Parliament.

The new Liberal government under W.L. Mackenzie King lost no time in attempting to co-opt the Progressives into the new cabinet. There can be little doubt that the impact of the agrarian party in Ottawa was shackled by the leadership's willingness to entertain some form of collusion with the Liberals over issues that advanced the agrarian cause. Through the Progressive leadership the Liberals worked to establish their influence over the former party. However, their efforts to co-opt the Progressives were frustrated by the protectionist wing of their own party and by the independent-mindedness of some prominent Progressive members. Among this group figured prominently Agnes Macphail, Canada's first woman MP and a UFO activist from Grey County, Ontario, who had also sided with J.J. Morrison against Drury's attempts to "broaden out" the farmers' movement in that province.[9] Despite the efforts of Macphail and others, the flirtations of the Progressive Party leadership with one of the old-line parties were no doubt significant in damaging its prospects of establishing a lasting political space on the Canadian scene.

Despite continued dissension in its ranks over its proximity to the Liberals, the Progressive Party managed to achieve some of its agenda. Although the protectionist faction within the Liberal Party was initially dominant, sustained pressure by the Progressive members eventually forced the Liberals into a reform of the tariff structure, simply to secure themselves in office. Progressive MPs also achieved a reinstatement of the Crow's Nest Pass Agreement and the lower freight rates it meant for grain shipped to port. Beyond this they could take much of the credit for the completion of the Hudson's Bay Railway to facilitate grain exports, and the construction of numerous rural branch-railway lines. They pushed through broader reforms that would have a benefit beyond the farm community as well, such as a new combines investigation act and conflict of interest legislation. To no small degree the Progressive movement represented an upswelling of regional resentment against the political economy enshrined in Macdonald's protectionist National Policy, and the perceived biases against rural hinterlands contained in that policy. After the Progressive upsurge, no federal party could afford to take the West for granted any longer. In addition to the transfers of some legislative powers from the federal level, the legacy of the Progressive movement meant that considerably greater care was taken from then on by Liberal and Conservative administrations to ensure that the West was better represented at the highest levels of government (Wood, 1975:362-63; Morton, 1950:288-90).

Finally, because of its radical wing the Progressive movement also brought about a push for political reform. In this endeavour it had less success, although it has been argued that its efforts were not entirely without results.[10] Its critique was nonetheless perceptive in important respects and bears a brief examination, especially since the flaws in the Canadian political process it sought to reform have endured to the present, with damaging consequences.

The Unrealized Dream of the Agrarian Movement:
Rejuvenating Canadian Democracy

While the agrarian foray into national politics did achieve a few concrete legislative victories, one very significant contribution has been largely lost with the tumultuous events of the intervening decades. It is worthwhile to try to recover this contribution, for its potential to enrich Canadian political life today is no less potent than it was seventy years ago.

Agnes Macphail

1890–1954

Teacher, politician, reformer, and feminist

Agnes Macphail broke new ground for women in many areas of the political arena in Canada: she was the first woman elected to the House of Commons and later one of two women elected in the Ontario legislature; the first woman to represent Canada internationally, as a delegate to the League of Nations assembly; and the first woman to sit on the League's disarmament committee.

She began her career as a rural schoolteacher in Bruce County, Ontario, close to her home, and continued teaching for ten years. While she enjoyed teaching, she found that it did not utilize all of her talents. She discovered that national politics was a cause she could devote herself to completely, especially using her talents as an organizer and dynamic speaker. She turned to the United Farmers of Ontario after the First World War, subscribing to a farmers' platform she could easily identify with and admiring the democratic principles and populist orientation of the organization. Macphail was elected to the House of Commons for the Progressive Party in 1921, the same year that Canadian women got the vote.

Macphail's political career also took her into provincial politics as an MPP for the CCF party in 1943 and again in 1948. Earlier she had helped build the alliance that became the Co-operative Commonwealth Federation (and later the New Democratic Party) and that finally ended the two-party system.

Macphail was a politician with reformer and feminist roots. Much of her work was aimed at improving human rights as well as women's rights. She was the founder of the Elizabeth Fry Society, which helped women in trouble with the law. She also worked for pay equity in Ontario in 1951, and as a feminist pacifist she participated in the Women's International League for Peace and Freedom. Macphail's brand of feminism was unique and timely. She was "an outspoken feminist committed to equality and human rights, [she] served as an essential bridge between first-wave feminism at the beginning of the twentieth century and its contemporary counterpart."

Her political career spanned three decades, and during that time she was elected eight times and defeated four times. In office she worked not only for farmers' rights but also for human rights, women's rights, penal reform, international peace, and pay equity.

Source: Crowley (1990).

The first dimension of this contribution is the powerful critique its radical wing made of Canadian parliamentary democracy. While virtually all components of the agrarian movement felt the prevailing two-party system failed to meet the needs of large segments of society, one wing of the movement developed a critique that challenged the very basis of the Canadian political process. This wing was especially strong within the United Farmers of Alberta under the influence of Henry Wise Wood. In the East, a strong element of the United Farmers of Ontario, represented most notably by J.J Morrison, could also be placed in this camp.

In their view, the dominance of Canadian politics by the two old-line parties, and the particular way these parties – when in power – organized the political process, served to limit the extent to which politics in Canada could function in a democratic manner. The agrarian radicals looked beyond the obvious corruption that was characteristic of Canadian politics at the time to the way in which the two-party system institutionalized autocratic government and thereby blocked a more genuinely democratic politics. The way party politics was constituted, they contended, prevented individual Members of Parliament from representing their constituents in any serious fashion. Instead, the party in power had come to be virtually controlled by a cabinet that had little accountability to rank-and-file Members of Parliament when shaping policy. Moreover, rank-and-file members had little chance of influencing the policy-formation process in cabinet. The same was not true of the most powerful financial and industrial interests, however. The agrarians felt that the business sector's bankrolling of the main parties at election time gave the companies privileged access to cabinet decision-making and enabled them to influence policy by circumventing formal democratic procedure.

To counter this reality of autocracy with its façade of democratic rhetoric, the agrarians fought for changes that would dramatically alter political life in this country. A first principle they agitated for was *constituency autonomy*, a concept implying that candidates should be nominated and supported by local electors rather than by a national party organization (Morton, 1950:119). Their goal was a representative who was "the delegate of [his/her] constituency, a moulder of legislation, and a critic of government" (ibid., 197). By the same token the Progressives expected a high degree of local accountability from their representatives, involving an ongoing contact between the elected representatives and his/her local electors. As the agrarian leader J.J. Morrison asked, "How can the consensus of the majority of the people be followed when a man is elected and does not see his electors for four years? Democratic government is impossible without contact. It is necessary to find out the will of the people" (quoted in Morton, 1950:121).

To ensure that representatives were responsive to their electors, and not merely passive adherents to the dictates of a national party caucus, the Canadian agrarians adopted an innovation from the U.S. progressive movement: the "recall." A number of Progressive candidates were required to sign a formal resignation – or recall – to be kept in the hands of the committee that organized the candidates' nominating convention. With this device in place, failure to be accountable to one's constituency could exact a high price. With this and other radical innovations the agrarians dared to remould Canadian political reality and make it a more thorough-going democracy. Just as at the economic level the agrarian movement fought for the principle of co-operation as against the unceasing competition offered by the capitalist market, in politics its members opposed the competitive partisanship and mock parliamentary warfare between the main-line parties that so often got in the way of serious debate of major issues; and they opposed the autocracy of the caucus rule within the traditional party system (Morton, 1950:92).

The high degree of independence and responsiveness to local concerns implied in the principle of constituency autonomy ran against the old-style party politics, and the near total subordination of parliamentary members to party policy formulated in cabinet and caucus. For the principle to work there would have to be a radical change in parliamentary politics. Indeed, some agrarian thinkers had already elaborated a blueprint for the type of changes needed. The radical agrarian wing was basically opposed to entering the political process as a *party* at all, because of their powerful mistrust of the old-style party-dominated politics. Henry Wood popularized the view that farmers should only enter the political arena as an *economic organization*: "In the permanent ties of a common economic interest were to be found that stable base which would prevent political action ending in disaster for political movement and association, as it had done for so many farmers' organizations in the past" (ibid.).

The radical agrarian wing was heavily influenced by Wood's ideas of government through organized economic-interest groups. This notion was based on an appreciation of the class nature of modern capitalist society and on the conviction that in such a society politics organized along traditional party lines inevitably allowed the most economically powerful class to maintain the upper hand. Wood's theory accepted the fact of class conflict in capitalist society and did not try to sweep it under the rug in the fashion of the Liberals and

Tories. Rather, his proposal sought to institutionalize this conflict in the political process.

In Wood's view, if each class were well organized, the exploitation of any interest and any individual would be minimized. As he stated:

> We go down there [to Edmonton] as farmers, we ask something we are not entitled to. The other classes are just as thoroughly organized as we, and they will resist any unjust demands, and that resistance of each other will eventually bring them to a common level, on which these great class differences will be settled, and they will never be settled in any other way ...
>
> Now, I know we are just as selfish, as individuals, as any other class, and I know that if we were organized as a class we would be inclined to do unjust things, and we are just as bad as any other people and they are just as bad as we are. But when every class is organized, and we come together and find these things will be a resisting force when another class tries to do something wrong, and it is only through the law of resistance that these things will be properly carried out. (Quoted in Morton, 1950:91)

Wood and like-minded thinkers, as Morton argues, were attempting to graft a program of guild socialism on the farmers' movement and a program of group representation on the political movement that was taking shape within the general agrarian revolt in Canada at that time. Moreover, the radical agrarians were moving away from the farmers' traditional acceptance of a free world market to embrace the need for state regulation and nationalization of key economic sectors of society. This view was undoubtedly shaped by the rapidly changing economic structure of the Canadian economy – the replacement of an economy based on the competition of small and medium firms by one in which combination and monopoly in many vital sectors had become the order of the day, precisely to avoid competitive pressures and its negative effect on profits.

Having found that the Canadian parliamentary system rarely allowed them to secure adequate representation, the radical agrarians advocated what amounted to radical reforms. Historian W.L. Morton describes the basic features of government that Alberta Progressives under Wood, and others in the agrarian movement, were struggling for. Their reforms advocated that:

> The elected members of each economic interest were to constitute a group in the legislature, which would be composed entirely of such

groups. In a legislature so constituted, the traditional organization of government and opposition parties would be impossible, with the passing of party government there would be an end of the domination of the legislature by the cabinet and of the cabinet by the prime minister, which, in the eyes of Albertan Progressives, were simply additional aspects of that party "autocracy" of which the caucus was the serious evil. Instead of such a party cabinet there would be a composite cabinet, made up through the proportional representation of the groups in the legislature, which was to hold office until deliberately dismissed by a vote of want of confidence in the legislature. (Morton 1950:150)

With the demise of the Progressive Party as a viable political force by the late 1920s, the thrust for radical reform empowered by the agrarian revolt suffered a major setback. This reforming impulse was not to die out entirely, however. The boom times of the 1920s were not to last, and when the economic cataclysm of the 1930s set in, the well-spring of much inspired agrarian thinking and organization, the Prairies, was to bring forth a more durable political response to confront an economic system that proved unable to meet the needs of the majority of agrarian producers.

CHAPTER THREE

THE CCF: A LASTING
POLITICAL VEHICLE

• • • • • • • •

*No CCF Government will rest content until it has
eradicated capitalism and put into operation the full
program of socialized planning which will lead to
the establishment in Canada of the Cooperative
Commonwealth.*

Closing paragraph, Regina Manifesto, 1933

The most incisive analysis of the weakness of the agrarian movement,
and the strategies that might bring it success, were both to be found among
the radical elements of the rural Prairie farming community. The experience with co-operativism on the Prairies demonstrated that in addition to an
understanding of their problems, the agrarians had an uncanny ability to
move from the realm of analysis to the realm of concrete implementation of
strategies for change. They saw the limitations of co-operatives in a context
in which economic power in society was concentrated in a few private
hands, and this pushed the radical leaders among them to propound a more
all-encompassing solution. Once again they turned to the idea of building a
political party – but this time, unlike the Progressive Party experience, it
would be a political party informed by a profound critique of the existing
order. It was an effort inspired by the objective to radically reform the existing social order, an effort that drew up just short of promoting a revolutionary break with the past. It was also a successful attempt to weld together,
once again, a farmer-labour alliance.

43

The Impact of Depression, Drought, and Agrarian Radicalism

During this time of transitory prosperity there were still a few agrarian leaders who refused to be lulled by the economic climate into a state of passivity. This was especially true on the Prairies. Despite the less than favourable national political climate, with the demise of the National Progressive Party the agrarian radical E.A. Partridge was actively criticizing the limitations of prairie capitalism and campaigning to replace Canada's capitalist economy with a co-operative commonwealth to end the "costly war of clashing vocational interests" (Partridge, 1926:80).

Other radicals were active in the rural scene of Saskatchewan in these years, patiently doing educational work through the farmer-owned Wheat Pool and the United Farmers of Canada (UFC) about the need for a new economic system as the only real solution to the problems affecting the rural community. By the late 1920s this agitation had led to a demand by the UFC leaders that the three provincial wheat pools take on 100 per cent of the wheat marketed in Canada, instead of the current 60 per cent level (Lipset, 1968:102). They felt this control was necessary if the producers were ever going to take the power of determining the price of wheat away from the speculators of the grain trade.[1] They urged legislation by the provincial government to bring this about.

This proposal was initially resisted. A powerful conservative group on the board of the Saskatchewan Wheat Pool favoured a policy of gradualism and opposed government intervention (see Fowke 1957:238-90). But this opposition, as well as some grassroots resistance, was to be washed away between 1929 and 1930 when the price of wheat went into a free fall from its high of around $1.50 a bushel. The price would eventually hit thirty-eight cents a bushel for No. 1 wheat in 1932 (Fairbairn, 1984:113) – and this was before transportation charges to the farmer were deducted.

Just as the tide began to move in favour of more all-encompassing co-operatives in the West, events at the international level were conspiring to undermine the foundation of the wheat pools themselves, and the central selling agency they operated to sell farmers wheat abroad. The collapse of the world wheat prices left the three pools in a dangerous financial situation, because their initial payments to farmers for the 1929 crop had been far higher than justified by the falling world price. With the banks clamouring at the door, the pools' leadership was scrambling to get concessions from the federal government to stave off disaster. By 1931 Western grain growers were dealt twin blows. First, the Saskatchewan Court of Appeal blocked legislation that would have given the Saskatchewan Wheat Pool control of all of the wheat marketed in the province. By this time as well the wheat pools had been forced, in return for federal assistance, to shut down their central selling agency's offices abroad and return to the much disliked system of selling through private middlemen. The pools' dramatic efforts to control the Canadian wheat trade on behalf of Western farmers had collapsed.[2]

In the words of a historian of the Western wheat pools, speaking of the Saskatchewan Wheat Pool in 1931, "Once, the Pool had been the core of a worldwide marketing organization whose day-to-day decisions were watched closely by businessmen in world capitals, whose selling strategy was important enough to bring a British cabinet minister to Winnipeg. Now, the Pool was a Saskatchewan elevator company struggling to cope with a $13-million debt" (Fairbairn, 1984:112-13).

The plight of the wheat pools and the collapse of their marketing agency brought into relief the radicals' warnings about the weakness of co-operative organizations as long as the wider economy continued to be organized along the lines of private market relations. The ethos of competition that permeated the wider society – even though in many sectors there was scarcely a semblance of competition – tended to frustrate the efforts of co-operative organizations to achieve the state of true co-operative enterprise for which they were established. In the end, without control of the "commanding heights" of the economy, in particular the financial system, co-operative enterprise would ultimately be held up to the dictates of the private banks and other institutions run on the profit motive. The radicals argued that without a more all-encompassing socialization of society, the "people first" principle of co-operativism would remain a utopian dream.

This message of the radical agrarians found a much more fertile ground among the farming population of the West as the twin forces of natural disaster (in the form of drought) and economic disaster (in the form of ruinous wheat prices) began to drive vast numbers of farmers into the hands of the banks and foreclosure. General statistical evidence for the time provides a glimpse of the magnitude of the disaster. Between 1928-29 and 1933, average per capita income fell by some 48 per cent for Canada as a whole; but in the largely farming economies of Saskatchewan and Alberta, incomes fell by 72 per cent and 61 per cent respectively (Lipset, 1968:122, Table 9). Table 1 shows clearly that compared to people receiving income from capital, such as bonds or stocks, and compared to small business or even to those receiving salaries and wages in protected industries, agricultural income declined so dramatically that it staggers the imagination.

Table 1
Decreases in Net Money Income by Source of Income:
1928-29 to 1932-33

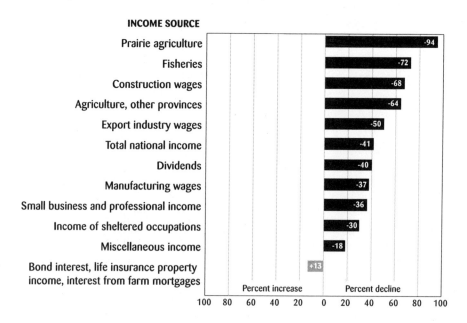

Source: Report of the Royal Commission on Dominion-Provincial Relations: Canada, 1867-1939 (1940), Book 1, p. 150, cited in Lipset (1968:123, Table 10).

The reality of disastrous wheat prices and drought, compounded by plagues of grasshoppers and the spread of wheat rust whenever the drought eased off, continued year after year on the Prairies. The economy did not bounce back in this pre-Keynesian era of "laissez-faire" state policy. Initially, at least, the federal Conservative government of R.B. Bennett continued the traditional policy of leaving the role of regenerating the economy to private capital, trusting that wages would eventually be forced down so far that there would once again be attractive investment opportunities. This was not at all surprising, because it was the policy favoured by big business itself, at least for a considerable time, and Bennett and his government had close ties to this group in society.[3]

However, this hypothetical shift towards full employment, this equilibrium towards which the market economy "naturally" gravitated according to neo-classical economic theory, remained a dream for most of the decade. Economic stagnation continued from year to year. By 1937 the situation was further accentuated in Saskatchewan by a particularly widespread crop failure. The average yield of wheat that year was only 2.6 bushels per acre – as compared to a ten-year average for the 1920s of 17 bushels per acre (Britnell and Fowke, 1962:72). In 1937 some two-thirds of the rural population were placed on government relief rolls (Britnell, 1939:97). The boxcars of food and clothing that private charitable organizations had been shipping to Saskatchewan for several years had to be massively supplemented by federal government shipments, and provision had to be made for the slaughter or removal of much of the province's livestock before it was destroyed by the drought. By the late 1930s the provincial government had become all but a ward of the federal government, so weak were the possibilities of raising revenues in the province. As deficits accumulated, outside borrowing from private sources became impossible (ibid., 99).

What did these statistics mean at "ground level"? George Britnell's contemporary account of the impact of the Depression on the farm population noted: "But since 1929 the standard of living obtaining under such conditions of relative prosperity has been replaced by a much more flexible depression standard, and under the pressure of low prices and short crops *this has been driven below what had previously been considered an irreducible minimum*" (Britnell, 1939:151, my emphasis). The depression of wheat prices affected practically all farms, but those in the southerly regions were hardest hit by drought, which lasted for years and resulted in much of the topsoil being blown away. As one farm wife recounted: "My second son was born in January, 1931, and that was the year we got no snow.... There was a tree outside my window that was budding out. It should have been twenty below outside.... That wind just blew and blew and blew and it just took the topsoil away. Miles high, so it seemed, into the sky" (Broadfoot, 1973:42).

Saskatchewan wheat farms typically supplied their families with 40 to 50 per cent of their food requirements straight from the fields. With the lack of rain year after year, this self-provisioning became virtually impossible for many: "Gardens were a failure. Potatoes will have to be brought in.... One man told of children whom he knew, who had not tasted any other vegetables than potatoes for over two years. Meat, bread and potatoes form the diet of the majority, it was stated, but this year it was feared that the meat will often be lacking and the potatoes as well" (quoted in Lipset, 1968:125). Farmers starved for cash incomes were forced to give up what were essentially basic necessities in the isolated circumstances of the Prairies – telephones and automobiles. Telephone subscriptions in the province fell by almost 50 per cent between 1930 and 1934; the number of cars registered fell by almost 40 per cent. The provincial school system began to decay as local school boards found it increasingly difficult to pay teachers' salaries and buy fuel in the winter. In any case, families were having great difficulties providing suitable clothes for their children to wear to school. By 1932 less than 40 per cent of the rural schools were open for the full year (Lipset, 1968:127-28).

The accumulated years of little or no cash income, of drought, plagues, and crop disease, had a heavy impact on the collective psyche, producing a continuing sense of desperation.

> The first calamity which we suffered was drought. It was a long torturous summer. Our grain, grass, fruit trees, and garden were all burned to the ground, leaving no feed for the horses, cows, and hogs, plus the fact that it left no fruit or vegetables to provide us for the long winter. We managed somehow for the first year, since it was the first time we'd ever had such an experience. We mended our old clothes, and resoled our shoes. We managed to buy food for ourselves and the stock and to pay the doctor and the dentist and the veterinary bill. It meant going without anything that wasn't absolutely necessary, and the first year we felt like experts....
>
> The following spring promised to be a good year, and optimists and gamblers that all farmers must be, they got seed on credit, and everyone worked hard and planted his crops.... Again the summer months became intensely hot and people were fearing what might happen. Then about noon one day everything became black. At first we couldn't imagine what was happening. Dust was sifting through the open doors and windows. Our priceless seeds were gradually being blown from the soil. I shall never forget that night as we sat silently eating our supper. It was happening to us again! Soon we would have no garden left. By this time we really had no money. Our only thread of life was our cows. There was enough grass and pasture that they could eat, and we sold cream for almost nothing.... We got through that winter somehow. We had to be on relief.... Again the spring seemed a bewitching promise. It was after this one that I came to loathe the springs. They were so deceptive, and I became a realist about the seasons. (quoted in Lipset, 1968:128-29)

If depression and drought provided the economic and social context for a renewal of the agrarian political presence, considerable credit must also go to the legacy of the earlier agrarian political movements. The role of the National Progressive Party was notable, in that it had helped to break down the rural allegiance to the two traditional parties of the Canadian political scene, a process that had been started during the days of the Union government of World War I. The Progressive Party was all but defunct by the mid-1920s, but its influence lingered in the minds of rural producers. When the movement had disintegrated in 1925, the influential Liberal John W. Dafoe, editor of the *Manitoba Free Press*, noted, "The Progressive movement represented a Western outlook, which has not vanished by any means. If it does not present itself through the media of the Progressives it will appear in some other form" (quoted in Sharp, 1971:186). The experience of the Progressive movement, however brief, had established the precedent of looking for a third party for remedial legislation, largely because of the deeply implanted belief that the old-line parties were corrupt and had been captured by the Eastern establishment (Sharp, 1971:189).[4]

Political scientist Walter Young argues that the core of the Progressive movement's legacy "was a heightened awareness of the feasibility of political action and of the importance of a distinctly radical platform to discourage the cannibalistic enthusiasms of Mackenzie King" and the Liberal Party (1985:540). However, the legacy of Prairie radicalism lies not so much with the Progressives, as Courville (1985) notes, but with independent farm and urban labour organizations that formed in the 1920s under the influence of a leadership largely socialist in its orientation. Their influences were not entirely limited to the province either, for agrarianism had spread its net across the country and felt a strong influence from the United States, most notably in the form of the Non-Partisan League. In Canada we should not forget the undeniable influence of radicals in the United Farmers of Alberta. Young describes those radicals as helping to generate a class consciousness among the rural population and as connecting the interests and grievances of labour to those of the farmers themselves. Then there was the influence of J.J. Morrison and the radical wing of the United Farmers of Ontario. In any case, once conditions became nasty on the Prairies, the door was already open for a "third way."

The Co-operative Commonwealth Federation

The emergence of the Co-operative Commonwealth Federation, or CCF, to become the governing party in Saskatchewan and a force to be reckoned with on the national political scene was another milestone in the development of agrarian political influence in Canadian society.[5]

The political developments in Saskatchewan in the 1930s were embedded in the province's political experience of ten years earlier. An important development had occurred in 1921, when a group of the most militant farmers, some with experience in the labour movement, had split off from the Saskatchewan Grain Growers' Association (SGGA). This group, which came to be called the Farmers' Union, had led the drive for the formation of the Saskatchewan Wheat Pool and been a constant critic of the close relationship of some of the leadership in the Grain Growers' Association with the Liberal Party in the province. By 1926 the Farmers' Union had fused with the SGGA to form the United Farmers of Canada, Saskatchewan section (Young, 1985:545). The group conceived of itself as a vanguard movement that would strive to steer the agrarian forces in a progressive direction through grassroots educational work (Lipset, 1968:88). This Farmer-Labour Group, as it came to be called, began the arduous work of grassroots organization and education around the province, assisted by the widely read weekly the *Western Producer*, the successor to the *Grain Growers' Guide*. The socialist elements of the new movement came to be widely accepted as leaders, most likely because of the failure of earlier strategies to advance farmers' interests, notably co-operative organizations limited to the economic level and attempts to influence the old-line ruling parties through political pressure (Lipset, 1968:135).

If one individual can be singled out as a catalyst in bringing together the key forces – the radicalizing farm organizations, the coalescing labour parties, and various social thinkers and politicians – it would have to be J.S. Woodsworth. A long-time socialist member of the federal Parliament from Winnipeg, this one-time preacher had for years travelled the Prairies and was intimate with the conditions, perspectives, and deep-seated problems of the Western urban working class and wheat farmers alike. He was, in short, a unique political figure in that he was trusted by both farmer and labourer, and by 1930 he had behind him a long and distinguished parliamentary record defending their causes. With views said to be more radical than many farmers and less radical than many labour party people, Woodsworth was "the linchpin between the urban labour groups and the farm groups, providing a central figure around which the party could unite" (Young, 1985:549-50). Coincidentally, the leadership of various labour parties in the West was pushing to co-ordinate efforts through a series of conferences bringing together the urban left-wing forces and providing a point of contact for the farmers' organizations (ibid., 547). These conferences were significant because they provided the opportunity to hammer out a labour platform to deal with the immediate economic crisis, as well as a program to confront the perceived inequities and injustices of Canada's capitalist economic system. Moreover, they provided a vital impetus for the formation of a national political vehicle to advance the policies being formulated. M.J. Coldwell, who would play a key role in the new party, remarked that these labour conferences deserved "much of the credit for the formation of the CCF" (Young, 1985:547).

One further ingredient in the mix that produced the CCF was the formation of the League for Social Reconstruction (LSR) in 1932. The idea for an organization that would bring together intellectual energies on the left to work for social change originated with two well-known progressive thinkers, Frank Underhill and Frank Scott (ibid., 552). Their intention was to fashion an organization, much like the British Fabian Society or the League for Industrial Democracy in the United States, that could provide the context for the development and dissemination of programmatic ideas that would in turn contribute to the development of a new class-oriented party (Horn, 1980:66). By 1933 the LSR had established branches across the country, although its energies tended to be provided by intellectuals, most of them university professors, based in Toronto and Montreal. While the LSR did not formally affiliate with the fledgling CCF, the two organizations shared a close relationship, with the key figures of the LSR providing an intellectual tutelage for the new party as it set out to distinguish itself from the two traditional parties, and the Liberals in particular (see Young, 1969:ch.4) The LSR gave a certain legitimacy to the CCF and to the idea of collectivist planning among teachers and other professional groups across Canada.

Political activity in Saskatchewan was given a further boost when likeminded movements in other provinces were convinced by Saskatchewan leaders to convene a convention to form a national movement. Meeting in Regina in 1933, the delegates decided to form a national political party, to be known as the Co-operative Commonwealth Federation (CCF). As Seymour Lipset notes in his classic study of the rise of the CCF:

> The Regina convention marked a turning point in the history of the Saskatchewan farmers' movement as well as in the socialist movement across Canada. For the first time the agrarians had become part of a national radical movement allied with urban labor. Instead of attacking only the aspects of the economy that could be controlled by cooperative action within the province, the farmers were now engaged in a frontal attack on the total economic structure. The organized class consciousness that had developed to defend the farmers' economic position now assumed the offensive. (1968:114)

J. S. Woodsworth

1874–1942

Minister, social worker, pacifist, and politician

James Shaver Woodsworth had an active career as a methodist minister, social worker, pacifist, and most notably socialist politician who was a founder of the CCF. Woodsworth was born in Ontario but moved to Brandon, Manitoba, in 1886 when his father received an appointment with the Methodist Church.

Woodsworth followed in his father's footsteps by becoming a minister. After high school he entered Wesley College in Winnipeg and studied in the Department of Mental and Moral Sciences. Following this joined the ministry and became a circuit-rider for two years in southwestern Manitoba. He also spent a further two years continuing his studies at Victoria College in Toronto and Oxford University in England.

His work experience in Canada and studies in England exposed him to the grim results of industrial capitalism, an exposure that made him question church beliefs and led him to conclude that the church's stress on personal salvation was wrong. Moving away from this orthodoxy, he began to expound what he referred to as the "social gospel." This was a movement calling for the establishment of the kingdom of God on earth. It was at this time that he became interested in social work and reform. He began to work in the slums of Winnipeg and traded his middle-class congregation for that of an inner-city mission.

The Regina Manifesto

At the first national convention of the CCF held in Regina, a manifesto with considerable input from the LSR was debated and ultimately passed by the delegates. In what came to be known as the "Regina Manifesto," the convention set down principles that were to guide the new political party and determine how members would strive to reshape the Canadian political reality. The Regina Manifesto is an extraordinary document, with its clear testimony concerning the degree of alienation from Canadian society as then organized and as experienced by a substantial part of the agrarian community. It also illustrated the preparedness of this agrarian community to embrace solutions

By 1914 he had become a union sympathizer and a believer in democratic socialism. He was also an impassioned pacifist who advocated in favour of anti-registration of men during World War I. These radical actions of openly opposing conscription cost him his government job as a director of research.

In 1918 Woodsworth resigned from the church because of its support for Canadian involvement in the war. To support his family he joined the longshoremen's union in British Columbia. A year later he was arrested in Winnipeg and charged with libel for an editorial he'd written about the Winnipeg General Strike. The case was eventually dropped, but he became better known for his work and soon plunged completely into politics. He began doing organizational work for the Manitoba Independent Labour Party, and in 1921 he was elected to the House of Commons, holding the seat until his death in 1942.

He collaborated with the radical Progressives, and when the Depression hit they joined with labour and socialist groups to found a federal socialist party in 1932: the Co-operative Commonwealth Federation. In 1933 the CCF met in Regina to create a manifesto based on democratic socialist principles. As well, they chose Woodsworth as their first leader. During the Second World War he strongly supported pacifism and tried to persuade the Canadian government to adopt neutrality. The CCF was split between supporting Woodsworth's pacifism or the country's entry into the war.

At the CCF's National Council in 1939 the party took the position of supporting the war, thus not adopting Woodsworth's pacifism. Woodsworth was alone in his advocation of pacifism, and the stand cost him his popularity. In 1940 he won his last election, but with a reduced majority. Two years later a severe stroke cost him his life.

Source: McNaught (1959).

that, at a minimum, constituted a radical reform of, if not a revolutionary break with, the old social order.[6]

A few passages of the Manifesto's introduction indicate the general direction intended for the CCF's political agenda. After opening with a statement that the CCF's purpose was to establish a Co-operative Commonwealth in which society would be organized for "the supplying of human needs and not the making of profits," the manifesto announced a more specific agenda: "We aim to replace the present capitalist system, with its inherent injustice and inhumanity, by a social order from which the domination and exploitation of one class by another will be eliminated, in which economic planning will su-

persede unregulated private enterprise and competition and in which genuine democratic self-government based on economic equality will be possible."[7] The document noted the glaring inequalities, waste, and instability of
the existing capitalist economy, and the extraordinary concentration of wealth
and political power within a "minority of financiers and industrialists." These
gross economic, social, and political deficiencies could only be rectified, the
Manifesto proclaimed, in "a planned and socialized economy in which our
national resources and the principal means of production and distribution are
owned, controlled and operated by the people." While stating "We do not
believe in change by violence," the Manifesto argued that the old-line parties
in Canada "are the instruments of the capitalist interests" and therefore "The
CCF aims at political power in order to put an end to this capitalist domination
of our political life."

After establishing in no uncertain terms the serious nature of its reform
agenda, the Manifesto set out in more detail the essential spheres in which
radical reforms were required. These included the areas of planning, financial
reforms, social ownership, agricultural reforms, reforms to the labour code,
socialization of health care, and a program to deal with the economic emergency caused by the Depression.

To establish a rational planning mechanism in the place of the vagaries of
the market, the Manifesto proposed a National Planning Commission to plan
production, distribution, and exchange of essential goods and services and to
co-ordinate the socialized industries. The Commission would also conduct
the research necessary for the efficient planning of the economy. Essentially
the desire was to replace the control of the leading sectors of industry and finance then in the hands of a few powerful capitalists with a staff of experts
supported by the necessary technical staff and public servants and accountable to elected representatives of the people.

Control of the financial affairs of the nation was to be removed from the
hands of an elite of private capitalists and placed under public control. This
entailed the establishment of a central bank to control the flow of credit and
regulate foreign exchange operations. The document also called for public
control of the private chartered banks, which had by the 1930s concentrated
the banking industry into only five companies. A national investment board
was to be set up to "direct unused surpluses of production for socially useful
purposes," while the socialization of insurance companies would allow for
some public control over this major source of public savings.

Under the rubric of "social ownership" the Manifesto set out an agenda for extending public ownership of the economy, first by incorporating utilities, transportation, and communications. This was a demand voiced in earlier Western agrarian programs such as the Farmers' Platform of 1917. However, the Manifesto extended this list to include mining, pulp and paper, and the distribution of milk, bread, coal, and gasoline. While advocating that "the welfare of the community must take precedence over the claims of private wealth," the architects of the Manifesto envisioned public enterprises managed along efficient economic lines and avoiding the rigidities of civil service rules and the evils of political patronage. They also called for a society where the employees in these enterprises "must be given the right to participate in the management of the industry."

While the changes called for in other areas of the economy were far-reaching, for agriculture the Manifesto was more cautious, accepting family farming as the basis for agricultural production and essentially calling for reforms that would make this more viable. The reforms entailed such demands as ensuring security of tenure on the land for farmers, without offering any great detail on how this would be achieved. Provisions such as crop insurance were mentioned, as well as state encouragement for the extension of consumer, processing, and marketing co-operative enterprise, and a system of import and export boards for farm products. Beyond this the document saw the improvement in the farmers' lot as being tied to the improvement of purchasing power in the wider society and the possibility this would bring for raising domestic agricultural prices. The absence of more concrete proposals most likely reflected the leading role of Eastern Canadian academics, more in touch with the problems of urban labour, in formulating the document.

In a couple of other vital areas – health care and the development of a labour code – the Regina Manifesto set out a clearer and uncompromisingly progressive statement. In its demand for "a properly organized system of public health services," stressing preventative rather than curative medicine, the Manifesto foresaw the revolution in health care in Canada for which the CCF was to serve as the vehicle. Much later, in the 1980s, demands for equal pay for work of equal value and "pay equity" to redress gender imbalances in salary structures would become a battle cry, but these issues were in fact raised in the Manifesto of 1933, with its call for "equal reward and equal opportunities of advancement for equal services, irrespective of sex." In its demands for unemployment insurance and social insurance to protect workers and families from sickness, industrial accidents, and old age, in its demands for limitations on working hours in accordance with technological developments, and in its call for unrestricted rights to freedom of association for workers, the Manifesto was calling for what today we understand and accept as much of the foundation for the modern welfare state.

Finally, assuming the CCF won power, the document saw the immediate need for the party to elaborate an emergency program to confront the critical economic situation. It envisioned a far-reaching program of public works entailing housing construction, slum clearance, hospitals, libraries, schools, community halls, parks, recreational projects, reforestation, and rural electrification – projects that would serve the twin purposes of getting the labour force back into productive and socially useful work and of building a new social and economic infrastructure for the next generation. This emergency program was viewed as a temporary measure; the Manifesto made clear that the real problem lay in replacing a defective socio-economic system. It closed with the statement: "No CCF Government will rest content until it has eradicated capitalism and put into operation the full program of socialized planning which will lead to the establishment in Canada of the Cooperative Commonwealth."

Electoral Victory and Implementation of the Farm Program

Shortly after the party was formed, it was thrown into its first election in the province of Saskatchewan. The 1934 election quickly became nasty, with the old-line parties attempting to scare the farmers away from the CCF. With the help of the sympathetic daily newspapers, the Conservatives and Liberals warned the farmers that they would lose their land in a CCF victory, because the party would call for blanket socialization of the economy. M.J. Coldwell, the CCF leader in Regina, was told to withdraw from politics or lose his job as school principal. He refused and was fired. On top of all this, the Catholic Church began a campaign to "warn" parishioners about the dangers of a CCF victory (Lipset, 1968:137). The fact that in the end the CCF achieved almost 30 per cent of the rural vote in the election, despite its lack of funds, a hostile press, and the absence of any sitting members in the legislature, was no small achievement. As Lipset (1968:140) notes, the CCF and the other emergent Western party – Social Credit – together managed to pull almost one-half of the rural electorate away from the Liberal and Conservatives.

Still, the leadership of the new party had expected more given the effects of the Depression on the farm community, and the 1934 defeat and the smaller vote they achieved in the federal election the following year led to serious soul-searching in the party. With shift in strategies, the party de-emphasized socialism and the nationalization of land in favour of more immediate and less far-reaching reforms. As well, it opted for a strategy of alliances with Social Credit and "progressive" Conservatives to defeat the Liberal government, a strategy opposed by Coldwell and Woodsworth. Students of this movement, such as Lipset, have argued that this change in strategy amounted to a significant turnaround for the CCF: "The whole purpose and program of the CCF had therefore changed. It could no longer carry on a systematic attack on capitalism and all its institutions. The party had become a farmers' pressure group seeking to win agrarian reforms" (Lipset, 1968:144-45).

A few years later, in 1938, in the midst of one of the worst drought-years of a drought-ridden decade, the CCF did surprisingly well in the Saskatchewan provincial election. Despite this, many CCFers pushed to repudiate the strategy of alliances with the Social Credit and sympathetic Conservatives, in favour of a "go it alone" policy. This revision back to their earlier position was to pay off in the provincial election of 1944. As it happened, the fortunes of the CCF more generally in the country were on the rise in the early 1940s. Membership was quickly increasing nationwide, and in the 1943 Ontario election the CCF gained a surprising thirty-four out of a total of ninety seats, and only four fewer than the victorious Conservatives. Federally it won four of the eight by-elections it contested between 1942 and 1945 (Zakuta, 1964:59).

Several factors account for this rise in CCF popularity. As one observer noted, "The Beveridge Plan, a comprehensive welfare scheme, and admiration for the Soviet Union were all in the air. Although the depression was over, its memory was still fresh, and people everywhere agreed, in language surprisingly like the CCF's, that the world must never return to its pre-war state" (Zakuta, 1964:58).[8] A major factor in the CCF's rising tide of success was the support it began to receive from a rapidly expanding urban unionized labour force. Under the determined efforts of David Lewis, an emerging CCF organizer and future leader whose orientation was towards trade unionism rather than the radical agrarianism of the Prairies, the party made links with powerful bastions of organized labour. Initially the leadership of the United Steelworkers of America was particularly supportive, soon to be followed by the Cape Breton local of the United Mineworkers union (Young, 1969:80-84). Further successes accumulated during the 1940s in building a broader base within organized labour. All of this began to change irrevocably the character of the CCF as a political movement and party, particularly outside the Prairie hinterlands.

Although the future of the CCF was on the rise generally in the 1940s, it was only in Saskatchewan that this popularity was actually translated into real political power. Under its new popular leader T.C. "Tommy" Douglas, the CCF managed an astounding victory in the provincial election of 1944. The incumbent Liberals won only five seats in the legislature, against the CCF's forty-seven. Social Credit managed only two seats (Silverstein, 1968:107). This gave the CCF a remarkably clear-cut mandate to initiate its reform program, and it went about this business over the next few years. The principal advances fashioned by the CCF victory came in legislation affecting agriculture, labour, and health care, priorities that reflected the social base – farmers and urban workers – that the CCF had welded together to achieve electoral victory.

The new CCF farm program was ushered in quickly after the election. Comprehensive and innovative, it broke new ground on several counts. For example, it brought in legislation to protect farmers from financial losses due to natural and other disasters, which were endemic especially with grain farmers. Farmers were given the legal right to retain enough of their crop until harvest "to keep families, pay unpaid harvesting and farm operation costs, and have sufficient seed for the next crop" (ibid., 114). The new Farm Security Act went so far as to ban mortgage foreclosure proceedings on the quarter-section of land containing a farmer's buildings. Under the act, in crop failure years farmers were also able to postpone interest payments and have their principal reduced by the equivalent of a year's interest. In terms of the usual relations governing the real-estate market, these changes were little short of revolutionary. Not surprisingly, they provoked a decidedly hostile reaction from the banking and loan companies.

The government pursued this reform vigorously, and the changes were reported to be well received by farmers generally. The reforms were not without problems, however, including shortages of funds, which placed limits on the program's eventual success (ibid., 116). In effect, through this legislation the CCF was attempting to make an important shift in the relative power of farmers and financial capital, in favour of the former. Other parts of the CCF farm program were less controversial, such as the establishment of seed and fodder banks as insurance against poor harvests, and the dramatic expansion of rural electrification.

One further development that benefited farmers especially – though not them alone – and helped to shift the economy away from a dependence on private capital was the establishment of a Department of Co-operation and Co-operative Development, which was to plan and co-ordinate government-sponsored co-operative activities. This occurred along with a substantial injection of funds to stimulate co-operative development (ibid., 117). There was even an attempt to promote co-operative farms to counter the growing inequalities among farmers and circumvent some of the irrational aspects of private farm ownership. This plan had limited success, in part because of the deeply ingrained individualistic ethos of the farm community and the pressures emanating from the wider, still basically capitalistic economy.[9]

Labour Reforms

In some ways the CCF's advances with respect to labour legislation brought the situation prevailing in Saskatchewan in line with other provinces with more advanced legislation, and in other ways they moved it ahead of other jurisdictions. The new Trade Union Act established labour's right to collective bargaining, as well as the compulsory check-off by unions that were recognized bargaining agents for workers, helping to redress the strong bias towards employers that had characterized provincial labour relations hitherto.[10] The Act went so far as to provide for punishment by expropriation in extreme cases where an employer's anti-labour practices constituted chronic violations of the new legislation. The CCF also raised the minimum wage substantially and introduced mandatory two-week vacations. Over the longer term, the opening of the door to unionization of the public-sector workers in the province was probably the CCF's most significant reform, especially in terms of its own future. Workers in the civil service and Crown corporations were encouraged to join unions and gained the closed shop and higher wages (Silverstein, 1968:124).

Typically, legislative reforms to the labour code are not especially meaningful unless there is the political will to enforce them. In the case of Saskatchewan, there is strong evidence that the CCF did back up its new labour legislation. For several years after the CCF election, the increase in union membership in Saskatchewan occurred at a higher rate than in any other province (Lipset, 1968:280, Table 39), and the workplace that witnessed some of the most dramatic increases in unionization was the public sector. By 1963 some 94 per cent of provincial employees were members of the Saskatchewan Government Employees Association (Silverstein, 1968:127). It was of no small significance that public-sector workers in the province associated their improved working conditions, enhanced security, higher wages, and expanded bargaining rights with the CCF – especially during a time when the state sector in capitalist countries was generally in a phase of long-term expansion. These reforms, it would seem, went a long way towards extending the CCF's social base of support in future years and thereby secured for the party a future in the province's political scene. In fact, even by the 1944 election and with only the *promise* of labour reforms, the CCF had dramatically improved its appeal among the urban working class, as compared to its initial electoral campaign ten years earlier.[11]

Initiatives to Create Public Enterprise

One of the less controversial but nevertheless substantial innovations made by the CCF was in the area of public enterprise. Through special legislation the new government put in place the machinery required to launch public corporations, either through the purchase of existing firms or the creation of new ones. The experience of non-capitalist forms of enterprise in Saskatchewan, such as the co-operative elevator companies and the Wheat Pool, opened the way for popular acceptance of this reform. So, too, did favourable public sentiment around the need to create indigenous enterprise, given the experience with large Eastern-controlled corporate capital throughout much of the province's brief history. Nevertheless, once in power the CCF moved much more cautiously on this front than the earlier leaders of the party, those who framed the Regina Manifesto of some ten years earlier, would have liked.

A particular area of innovation was public insurance. Pointing out that private insurance companies were draining the economy of funds that were typically invested elsewhere with little benefit to the province, the CCF promised a non-profit government insurance scheme that would return people's money back to them in the form of compensation payments (Silverstein, 1968:137). Out of this initiative came public auto insurance, which provided a valuable model for other Western provinces in their attempts to set up successful public schemes in the postwar period.

The Push for a Universal Medical Plan

Perhaps the program that is most closely associated with the CCF's longer-term legacy to Canadian society was the fight to introduce universal health care, or what some have referred to as "socialized medicine." The CCF's eventual success in this struggle set the example for a national health scheme in Canada, arguably a scheme that especially in recent years and more than any other aspect of our society has come to distinguish us from the model of society presented by the United States.

The gist of the plan was to provide every citizen with adequate medical care irrespective of financial circumstances. To achieve this goal the party introduced a plan in the 1944 electoral campaign that called for provincial physicians to become salaried employees of the provincial government, for the discontinuing of private practice, and for the Department of Health to be given ultimate supervisory control over provincial medical practice (Silverstein, 1968:138). Doctors in the province fought the proposal "tooth and nail." Threatened by the loss of some of their traditional privileges and the almost total control they and their professional association exercised over health care, doctors rapidly organized and enlisted the opposition Liberals and the largely Liberal daily press to advocate their cause. In the press they repeatedly threatened to migrate *en masse* out of the province if the CCF plan were to be implemented.

For a long time to come this concerted opposition by physicians together with a lack of federal financial support would block the implementation of the universal health-care scheme. In its first term in power, the CCF had to retreat to a fall-back position on health care. Although the plan for universal medical insurance was stalled, the party was able to implement a provincial hospital plan that provided a virtually complete range of hospital services as benefits with no limitations in entitlement days as long as in-patient care was medically necessary (Taylor, 1978:102). It also passed legislation to provide free medical and dental care to all persons receiving welfare, including the blind, the aged, and those on mother's allowance (Silverstein, 1968:146). This was not an insignificant achievement, considering that at the time few people could afford to use hospitals except in the direst emergencies. Largely because of this situation, by 1940 in some provinces the majority of births still took place outside of hospitals. Infant mortality rates ranged from thirty-eight out of every thousand live births in British Columbia to seventy and eighty per one thousand live births in Quebec and New Brunswick respectively, a dramatic difference compared to the rate of around eight per one thousand live births for Canada as a whole by the mid-1980s (see Taylor, 1978:5; World Bank, 1988:287).

This CCF initiative of providing the first hospital legislation in North America left a legacy to the rest of the country as well. As one well-known work on health insurance and public policy in Canada notes: "All of Canada benefitted from the Saskatchewan experience. From its inception until the Quebec Hospital Plan came into effect in 1961, no provincial government failed to send its officials to Regina to learn at first hand how the program operated, and what policies and procedures could be adapted to their home provinces. In the educational process through which Canadian governments learned how to administer universal hospital insurance, Saskatchewan paid most of the tuition fees" (Taylor, 1978:104).

By 1960 the CCF, still headed by T.C. Douglas and still in power, decided that the time had come to push through the full medical insurance scheme. In the election of that year the medical issue was a central campaign theme. The College of Physicians and Surgeons organized a vigorous opposition campaign, enlisting support from the provincial Liberal Party, the Pharmaceutical Association, and the Chamber of Commerce. They also had the support of the press, much of it in Saskatchewan owned by the powerful Sifton family, which had strong ties to the opposition Liberals (Naylor, 1986:184). The physicians put in place a well-organized public relations campaign against the proposed plan, using the tactics developed by the American Medical Association, which had earlier waged a rather rabid campaign against medicare in the United States. The campaign organizers sent leaflets to every household in the province outlining the supposed "dangers" of the CCF plan. The doctors in Saskatchewan could also count on the support, monetary and otherwise, of the Canadian Medical Association and the Ontario Medical Association in their efforts to defeat the plan.

These efforts very nearly worked. The CCF won the election of 1960, but by a narrow margin. The well-organized physicians, with their external support and backed by a sympathetic press, opted to continue their intransigence to the plan. Negotiations stalled, and as the government moved to implement the plan the doctors threatened to withdraw services. By the early summer of 1962 the province was moving rapidly towards a doctors' strike. The physicians clearly had powerful allies in the business community, but also among other professional groups and all those who could be rallied by a call against state intervention. The CCF government, for its part, was supported by the provincial labour federation and a substantial proportion of the farm community. The health issue was rapidly polarizing the entire province along ideological and socio-economic lines (Naylor, 1986:203) and had become a national issue of the first order.

By early July the majority of the doctors in the province had closed their offices (ibid., 207) and a degree of hysteria was being fostered by the provincial press. The government compromised as far as it felt it could without losing control of the essentials of the plan. The intransigent attitude of the College as the negotiating agent for the doctors, and the doctors' decision to withhold services in particular, brought on approbation and even open hostility in much of the rest of the country.[12] The issue became internationally recognized as well, with important elements of the British medical establishment condemning the withdrawal of services, while the more right-wing and pro-

"free enterprise" medical groups in the United States lending support and donations (ibid., 207-8).

In the end the anti-government campaign waged by the doctors and their sympathizers began to lose support, in part, no doubt, because their move to withdraw services undercut the moral authority of their cause. Partly through the mediation of a well-known British physician and Labour MP, Stephen Taylor, who had helped set up the British National Health Service, a compromise was finally agreed upon after much delay and failed attempts. The doctors were clearly looking to the next election, in which the Liberals had a good chance of winning, to bring an end to the plan. The Liberals did, in fact, defeat the CCF in the elections of 1964, but they proved unwilling to dismantle the popular medical insurance scheme.

The comprehensive medical plan brought in by the CCF covered all medical services except dental and optical care. Although citizens were generally expected to pay a monthly premium, nobody was denied access to the plan because of inability to pay (ibid., 195). Participation in the plan was compulsory for all licensed physicians, who were paid from government funds. The government set up an advisory committee, with representation from the provincial medical association and municipal representatives, to manage the implementation of the plan and iron out administrative problems. The major decisions over policy were kept under the purview of the elected members of the legislature (ibid., 194-95).

With its victory in the struggle over health insurance, the CCF had managed to bring into being the first universally tax-supported medical insurance plan administered by a public authority in North America. This singular victory is credited with having moved the matter of public-health insurance up to be a matter of first priority on the political agenda of the federal government and all the provinces (see Taylor, 1978:328).[13]

Final Reflections on the CCF and Its Achievements

Viewed today, the aims and actual legislation of the CCF might seem to have been misguided – steering us in the direction of the modern welfare state, which we are being told is increasingly unviable and inefficient. There are at least two problems with this view. Firstly, it does not examine critically enough the real economic interests that are now being served by the much touted swing in sentiments away from state intervention towards a "market-oriented" approach. Secondly, it divorces the struggles of the agrarian and labour forces behind the CCF from the actual historical and socio-economic context in which they took place, and which made these struggles appear to be highly sensible, that is, both realistic and necessary in the eyes of those favouring them.

The first issue is a very large one and merits a study all its own. In effect, it involves the shift in dominant ideology affected since the 1980s through a number of dimensions that generally impinge on how our society is organized. At the core of the ideological struggle, for that is what it has been both in Canada and in much of the rest of the world, is a debate about the appropriate role of the state and public enterprise versus the role of private capital in organizing our economic lives and our society. The disintegration of the centrally planned economies in the Eastern bloc countries, including their headlong rush towards capitalist enterprise and the substitution of planning by "market forces," has been used by the private media conglomerates that control our information to strongly reinforce the notion that public enterprise is at best inefficient, and at worst inherently evil. It must be remembered, however, that the most powerful elements in our business community both in Canada and the United States were lobbying hard for the massive changes affecting the relationship between state and society for much of the last decade.

These changes include the deregulation and downsizing of the public domain through massive sell-offs of public enterprise to the private sector. Nor have the largest corporations that control our economy been insensitive to the need to legitimate their actions in the public mind. The 1980s witnessed a growing interest by large corporations to fund private "think-tanks" in the form of institutes and research foundations. Staffed by "experts" who share the common view that private capital and the market are the best way to go, these organizations have played a key role in shifting public consciousness away from any sympathy for state involvement in the organization of the economy. To be sure, these orchestrated efforts would have been less successful if in fact serious contradictions and limitations had not become apparent in the mixed economy form of capitalism that emerged after World War II.

Nevertheless, over the last decade or so the way we have interpreted these contradictions has been assiduously shaped by powerful economic interests. Moreover, the debate on the solutions to the crisis of the welfare state in Canada and other advanced capitalist countries has been characteristically narrow and one-sided. Deregulation, privatization, and trade liberalization have to be understood within the context of the changing needs of multinational firms in the last decade of the twentieth century. The great danger is that while in the 1960s we used to question whether "what is good for General Motors is good for America," in the 1990s there is a great reluctance to ask whether what is good for General Motors today, with its Japanese-affiliated corporations and worldwide corporate operations, is also good for local citizens. Public enterprise at least had national and often regional concerns as part of its mandate, whereas the global corporations have no such concerns and loyalties.

The reality is that state intervention came about as a result of the pressures of organized interests from below, fighting in opposition to powerful corporate interests to try to deal with an economic system that was manifestly not serving the interests of the great masses of people, or even the interests of the chief benefactors of the system. The myopia of the present fails to perceive the problem inherent in the capitalist economy. The debacle of the U.S.S.R. and its allies only reinforces this myopia; it does not in any way dissolve the contradictions inherent in capitalism – contradictions that brought on the agrarian protests of more than half a century ago, and the eventual reforms.

The reforms finally carried through by the CCF in Saskatchewan can be seen as one outcome, and a generally positive one, of the decades-long impulse of agrarianism that had stamped the life of the emergent Canadian nation. Earlier impulses, such as the Patrons of Husbandry and the Dominion Grange, had lacked sufficient direction and purpose, or, as with the Progressive movement, were marked by fatal flaws in conception and execution; but to a large extent the CCF overcame these earlier flaws. Building on the initial insights of the Prairie radical agrarian tradition, the CCF was well inoculated against the dangers of co-optation by a reformist Liberal Party. The limitations of more than two decades of co-operativism instilled the conviction that more all-encompassing political solutions were necessary. Finally, the CCF was able to avoid the ultimately sterile clash that had marred earlier agrarian struggles: the conflict between those who would protect the purity of the movement's agrarian roots, and those who wished to broaden the movement's class base.

The CCF was able to forge an alliance of farmer and labour organizations, which was a workable and successful alliance because it had already constituted itself as an ideologically distinct movement, and in doing this it could integrate the progressive elements of both labourer and farmer under a distinct set of principles and a unique political agenda. Unlike the less ideologically secure Progressive movement at an earlier time, it was not in danger of being absorbed by an expansive and reformist Liberal Party.

CHAPTER FOUR

THE INSTITUTIONALIZATION
OF AGRARIAN POWER

• • • • • • • •

I am afraid this legislation [to reduce the guaranteed
price of wheat] if it carries will cost us many seats
in Western Canada.... It will be, I fear, a sort of
suicide to proceed with it.
Prime Minister Mackenzie King, 1939

We're saying the milk producers in Ontario are going
to get a stronger collective countervailing power in the
market place. We're going to get improved prices for milk.
George McLaughlin, Ontario Milk Marketing Board

Although the farming communities' thrust for political power on their own
had largely disintegrated by the mid-1920s, it was quickly replaced by the producers' struggle to gain control over impersonal market forces and resist the
rapidly growing clout of the massive new industrial and commercial enterprises of the twentieth century. This resistance and struggle to control the
chaos of a market-based economy brought a move to institutionalize their own
power – a process that would prove to be tremendously important in shaping
the post-World War II food system.

This process of institutionalization included the establishment of the Canadian Wheat Board and the development of a regulatory regime in the dairy
industry in Ontario – two examples we will use to bring into focus the different concerns of producers in the West and the East of Canada. These two
cases provide, in addition, different insights into the processes taken by pro-

ducers to advance their interests. The case of the Wheat Board illustrates that although producer pressure was undoubtedly a necessary precondition for institutionalization, it was not enough by itself to force state intervention to protect producer interests. That case points to the capitalist sector's role of acquiescence, and even support, in a time of acute crisis, together with external exigencies created by wartime conditions, as essential factors in explaining state intervention. The case of the regulatory regime for dairy producers in Ontario suggests how the state can play a relatively autonomous role, in certain situations, in promoting far-reaching legislation that benefits the general long-term interests of producers, processors, and consumers – even when particular producer and processor groups were preoccupied with their short-term objectives and adamantly opposed to change.

Amidst the chaotic market conditions of Depression-era capitalism, with their particularly drastic toll on Canadian agriculture, the objective interests of producers and agricultural processing companies began to coincide as never before. This coming together of interests involved a shared perspective on the needs for a general increase in agricultural prices and for ending the volatility and destructive competition that threatened producers and processors alike. The producers' interest in some form of price relief is obvious enough, given the fall in agricultural prices – a fall of drastic proportions in the case of wheat. What has been less obvious is the business sector's desire for state intervention to stabilize the troubled agricultural scene; more specifically, significant business interests in the processing industry, and also the banking establishment, shared an interest in rescuing the wheat economy.

As Alvin Finkel (1979) argues in his study of these years, the business community began to see government intervention as the only way to achieve two key objectives: first, to satisfy the need to pacify consumers and producers disrupted by the economic chaos; and second, to provide regulation that would stabilize an industry in danger of disintegration. For processors a key problem of the period was the overproduction of agricultural commodities, which threatened their own profitability. The more far-sighted members of the business establishment looked for ways to establish state-regulated agencies without giving farmers control over the prices and profits of distributors and processors (Finkel, 1979:48).

Nevertheless, the most powerful initial thrust for state intervention came from the farmers and their organizations. The severity of the Depression for the agrarian sector built up tremendous political pressure on the federal and provincial governments: they had to be seen to be addressing the problem. One manifestation of this pressure was the establishment in 1934 of a Royal Commission on Price Spreads. The Commission was to take a hard look at the whole food economy – from input manufacturers to farmers through to processors and retailers of food – and examine who was being protected and who was not, and why, from the economic crisis. In its 1937 report it documented the remarkably high degree of concentration in a number of important industries related to agriculture and provided copious evidence of the negative impact of this concentration for the farm sector.

At more or less the same time, considerable agitation was being carried out by farm organizations, with the objective of securing what came to be called "orderly marketing" – essentially a state-regulated alternative to what they saw as the destructive competition of the unfettered market. Orderly marketing is a concept that involves policies designed to achieve the maintenance of fair prices, the regulation of supply, the cultivation of demand, and the improvement of quality (MacPherson, 1979). Concretely, these policies entailed the creation of marketing boards with powers legislated by government, and a strong precedent for this had already been set in Canada during World War I in the grains sector. The fruit of this agitation was the National Products Marketing Act of 1934, which was passed in the dying months of the Bennett government. Finkel notes the significance of this legislation:

> [It] allowed a majority of producers of a primary product to establish rules for orderly marketing under government supervision. These rules would be put into effect through marketing boards established for the various primary products, and in turn a Dominion Marketing board was to be set up to approve and regulate the work of the various marketing boards. Marketing would continue through existing channels and no board would participate directly in marketing. Through their boards producers would be able to determine minimum prices, quotas, quality grades and other regulations deemed necessary. (1979:51)

While this was an important step in the right direction for the farm organizations, it also served the interests of the processors and wholesalers. According to the Conservative Party's Minister of Agriculture, Robert Weir, the aim of the legislation was to provide the necessary stability to allow the agribusiness sector to have a "uniform quality" at its disposal and to guarantee set returns to producers (see ibid., 51). The agribusiness community, or at least substantial parts of it, saw the stabilization that could be provided by marketing boards as a generally good thing, and they actually intervened in the policy-making process to ensure that legislation would reflect as much as possible their own economic interests. The final legislation, for example, prohibited marketing boards from being involved directly in the marketing of a product, an activity that agribusiness was keen to keep to itself (ibid., 53).

The push for a more active state involvement in agriculture suffered a setback with the election of the Mackenzie King Liberals in 1935. Soon after, the Judicial Committee of the Privy Council declared the Bennett government's National Products Marketing Act unconstitutional (MacPherson, 1979). Nevertheless, the forces that had agitated for the federal legislation were not about to let their efforts be wasted, and soon afterwards they pushed province after province to follow the aborted federal lead and implement marketing schemes at the provincial level.

In the meantime, by the late 1930s farm organizations were desperately trying to regroup themselves to advance the cause of their members with a less than sympathetic federal government. Two landmark conferences were organized, one under the auspices of the Premier of Manitoba, John Bracken in Winnipeg, and the other organized at Montreal by a one-time United Farmers of Ontario leader, H.H. Hannan, for Eastern farm organizations. These conferences proved to have a significant impact in advancing the agrarian cause, just as conferences had helped pull together the CCF some years earlier. In particular they served to focus farm pressure at the federal level to secure a comprehensive and centralized agricultural marketing program (MacPherson, 1979).

The struggle by the agricultural producers to institutionalize structures that could protect their interests and reinforce their power vis-à-vis other actors in the food economy had a number of dimensions. Two of the most significant, in terms of the number of producers affected and thereby their impact on the wider society, were the struggle to establish the Canadian Wheat Board and the concerted efforts to bring about state regulation of the dairy industry in Ontario. The establishment of the Canadian Wheat Board had a predominantly Western impact, while the consequences of regulation of the Ontario dairy industry were particularly important for Eastern Canadian agriculture.

The Struggle for the Canadian Wheat Board

Since the days of E.A. Partridge's hostile interventions with the Winnipeg Grain Exchange, Western grain growers had agitated for an alternative to the speculative system upheld by the grain traders. The defeat of the plans for the 100 per cent wheat pool in 1931 did not mean that Western producers had given up on the cause. It really only meant a shift back to a strategy of lobbying for a marketing board for wheat (see Fowke, 1957:262). The main weakness of this strategy was that it was dependent for its success on a number of factors over which producers had limited, if any, influence. Nevertheless, with the pools in a state of bankruptcy, attractive options were few, and the precedent for a national wheat board had already been set during World War I.

The process establishing the Canadian Wheat Board and the story of the Board's fate until the mid-1940s are complex. What interests us here is not the mass of details that characterized this significant step towards state intervention in the food economy, but rather a consideration of the interplay of social forces that actually brought it about. Here it is interesting to contrast Vernon Fowke's now classic interpretation in his *The National Policy and the Wheat Economy* (1957) with the work of a newer generation of historians of the food economy, exemplified by Alvin Finkel's *Business and Social Reform in the Thirties* (1979).

Fowke's careful study is not primarily concerned with the actual process by which the Board came into being and by which its powers were subsequently shaped and limited. He does lay emphasis, however, on the sustained pressure by the organized grain growers to elaborate an alternative to the privately controlled grain trade. Moreover, Fowke's interpretation tends to give credit to the primary producers themselves as the force that ultimately pushed through the Wheat Board plan with a reluctant federal Conservative government. There is little mention of the other forces that came to bear on the proposal. In Fowke's words, "The Conservative Government under Prime Minister Bennett eventually yielded to the representations of Western wheat growers who persistently urged the restoration of a national wheat board" (1957:263). He does note that the need to shore up Western support in the upcoming general election also played a role.

Finkel is much more concerned with the process of the Board's establishment, that is, with the interplay of forces that determined a given result. He points out that key figures in the Bennett government, including Bennett himself and John McFarland, whom Bennett had appointed to oversee Canadian grain marketing when the wheat pools collapsed, both had close ties to the grain-elevator and trading companies and were therefore less than sympathetic to the idea of a wheat marketing board. Until the mid-1930s, the government's position continued to be that the private grain trade, which included speculators who bought and sold futures in wheat, was the best system for marketing the nation's number one agricultural commodity. The farmers were against the speculation in futures because they saw that they typically ended up competing with each other by selling cheaply to futures traders to avoid being left with unsold wheat after the harvest. As Finkel notes, only a government agency with fixed prices and quotas for wheat could prevent the farmers from competing against each other to the speculators' advantage (1976:61).

But after 1931, as prices spiralled downward, the private grain market began to break down with the virtual disappearance of speculators. To prevent the total collapse of the grain industry, McFarland as the government's representative ordered massive government purchases of wheat in an attempt to stabilize the market. By the mid-1930s it was clear that some sort of government intervention was necessary with the almost total failure of the market to provide conditions for the continuance of Western agriculture. As McFarland noted in a letter to Prime Minister Bennett, "It is now quite evident that without the government support in these years, the Futures System would have failed in its essential function" (quoted in Wilson, 1978:462).

By this time leading businessmen in the grain industry, formerly "free enterprisers" to the last, began to back the idea of government support for the grain market. The major Canadian banks, too, whose main concerns were their loans to the pools and the Western provincial governments, proved to be even more solidly behind state intervention in the grain trade (Finkel, 1979:64).

With the failure of the futures market to exercise its traditional function, the grain industry, if it was to avoid drastic decline, was dependent for its survival on some form of state intervention – which had already occurred in most other wheat producing countries. To the grain producers' constant pressure for a national marketing agency, then, was added the acquiescence and even the active support of some sectors of the grain-trading and grain-elevator business. To this must be added the conversion of key government personnel, in particular John McFarland, who had lost confidence in the private market's ability to do the job.[1]

This confluence of events would seem to be responsible for turning the tide in favour of significant government intervention in a major sector of Canadian agriculture. It was a decision that, as Fowke has remarked, was to have an impact on the whole evolution of agricultural policy in Canada (1957:265). As originally submitted to Parliament, the bill to create the Canadian Wheat Board gave the Board powers to take over all grain elevators in the Prairie provinces and to exercise exclusive control over the movement of grain between provinces and for export overseas. It was to exercise its powers over coarse grains as well as wheat, and it was to be of a permanent nature, rather than a temporary device to deal with the existing economic crisis (ibid., 264). What happened to the bill provides an interesting lesson on the relative power of the various forces at play on this issue.

The grain company interests, originally in favour of some form of state support for the grain industry, came out steadfastly opposed to the idea of a compulsory board. Led by such powerful grain capitalists as James Richardson, they not surprisingly opposed what they viewed as "the practical confiscation of grain elevators in Canada." They also opposed all the compulsory clauses of the bill, arguing that it would prejudice Canadian interests "to lose the highly trained personnel of the Canadian grain trade" (Richardson, quoted in Wilson, 1978:471). The private grain interests wanted to substantially dilute the bill, but they were still favouring some kind of state support to stabilize the industry.

For their part the grain growers, represented by the leaders of the co-operative elevator organizations, lobbied in Ottawa to keep the bill as it was, but in the end most of the sections of the bill that the grain business found offensive were removed. The amended bill, as one well-known study of the Canadian wheat economy notes, "was a victory, in fact, for the [grain] exchange" (Wilson, 1978:473). What type of legislation did all this struggle and lobbying finally produce? It was apparently not what the wheat growers wanted, but rather reflected the effect of successful business lobbying. As Fowke notes, the amended legislation did not end the activities of the speculative futures market but only provided producers with an optional marketing channel. It thereby freed them from a dependence on the open market system, without interfering with that system (1957:265). The compulsory features of the legislation were not put into effect, as per the grain trade's wishes.

> The board was merely required to establish annually a minimum price at which it would purchase wheat offered for sale by the grower, and to issue participation certificates which would entitle him to share in any additional proceedings. The individual grower might sell all, or none, or any intermediate portion of his wheat crop to the board. The board's fixed price provided a floor below which no grower needed to dispose of his crop. If sales by the board yielded returns above the price paid initially to the grower, the surplus was to be distributed on a pooled basis, grade by grade. If sales yielded less than the initial advance ... the shortage became a charge upon the federal treasury. (Ibid., 265)

The election of the Liberals brought in a government that was more ideologically disposed to let the speculative futures market prevail and to view the Wheat Board as only a temporary measure. One of the Board's initial opponents was T.A. Crerar, one-time leader of the Progressive Party and now a Western Liberal cabinet minister (see Finkel, 1979:70). If Crerar was at odds with the more radical wing of the agrarian movement in the 1920s, the return of the Liberals found him out of step with almost all of the producer organizations, and even elements of the Prairie business community.

Under the Liberals the Wheat Board set a floor price for wheat low enough to guarantee that their allies in the grain exchange would be marketing most of the wheat. It was not until 1939, however, that the private enterprise bias of the King government with respect to the wheat economy came to the fore. The Liberals began that year with what Fowke refers to as "the settled intention of the government in the early weeks of the session to rid itself once and for all ... of the incubus of responsibility for the disposal of western wheat" (1957:268). In a major policy speech the Minister of Agriculture, Jimmy Gardiner, proclaimed that the government intended to follow the guidelines of the recent Royal Commission on the Grain Trade, which included the recommendation, "The government should remain out of the grain trade and our wheat should be marketed by means of the futures market system." More generally, he noted, "We cannot agree that there is likely to be any permanency to any system of marketing farm products which is based upon price fixing" (quoted in Fowke, 1957:269-70). With these pronouncements the Liberals seemed to be announcing a major retreat for the federal state in the area of agricultural policy, and in particular with what was arguably its most significant initiative to that time, the Wheat Board. What occurred in the short period afterwards provides important insights into the whole policy-formation process with respect to Canadian agriculture.

The government fixed the 1939 floor price of wheat at sixty cents, a price lower than previous years and completely out of step with what producers were getting in other countries.[2] This step, along with its pronouncements on marketing boards, served not only to coalesce primary producers in the West and the East, but also to bring business interests benefiting from the wheat economy to the producers' side. The two major agricultural conferences held in the West and East served to focus the agrarian protest around the Liberals' perceived intention to retreat on existing marketing board initiatives.

Delegations from Western farm and business organizations descended upon Ottawa, Prairie legislatures forwarded resolutions on the issue, and a wave of communications and petitions engulfed Members of Parliament, including one petition signed by 155,000 farmers for retention of the eighty-cent floor price for wheat. In the end the Liberals backed off their plans to jettison the Wheat Board and compromised somewhat on the initial payment by raising it to seventy cents a bushel. Prominent Western cabinet ministers, including Crerar, had warned that plans to undermine the Board and substantially lower the floor price for wheat would result in a considerable loss of seats for the Liberals in the upcoming election, and these cautions undoubtedly had an impact. Crerar also noted the broad base of the protest in a letter to

Prime Minister Mackenzie King: "The price matter is not one in which alone the farmers are interested, but the whole business community as well" (quoted in Wilson, 1978:598). Moreover, the Liberals included with the Wheat Board bill a plan to establish a crop-failure assistance program in an attempt to equalize the way government assistance benefited grain producers.

What did this reversal signify? One noted observer of the Canadian grain trade summarized its impact: "The really significant fact was that the Government had tried to get out of the business of selling wheat and had failed. Its inability to divest itself of the Canadian Wheat Board at this time registered the fact that the Canadian wheat growers were determined that the Government should maintain the Board, if not primarily as a regular vehicle for selling wheat, at least as a stand-by organization that would protect them from drastic downward surveys of the market" (D.A. MacGibbon, quoted in Fowke, 1957:274).

The reversal of the King government on the Canadian Wheat Board question as a result of the concerted storm of Western protest had more than a temporary significance, according to observers of the role of the wheat economy in Canadian society. As one of these remarked, "It was no longer possible to regard the Canadian Wheat Board as a temporary organization established to meet a single emergency." On the contrary, the Board had become "a permanent part of the international machinery of Canada for dealing with wheat marketing" (quoted in Fowke, 1957:274).

Indeed, not long afterwards, in the midst of the Second World War, an entirely different set of circumstances than those prevailing during much of the 1930s forced the same government to move from a voluntary to a compulsory Wheat Board. The Board then became the exclusive initial recipient of Canadian wheat delivered from the farm, and the trading of wheat futures on the Winnipeg Grain Exchange ceased entirely. There were wartime reasons for this final move: compulsory state marketing was initiated in 1943 to ensure that wheat would be available to continue the delivery of large bulk sales of wheat to Britain established at the beginning of the war, and to fulfil obligations Canada had assumed under Mutual Aid agreements with other countries. This arrangement was endangered by the rapid advance of wheat prices in the open market in 1943, which was in part occasioned by a strong demand for Canadian wheat developing in the United States and also by a shortfall in the 1943 wheat crop (Britnell and Fowke, 1962:214-15). It is worthy of note, especially given the negative attitude towards marketing boards that is now fashionable in certain circles, that this entrenchment of a significant marketing board arrangement came not as a means of propping up prices for farmers in a desperate situation, but as an attempt to prevent open market prices from escalating drastically and thereby jeopardizing the federal government's wartime commitments.

The Significance of the Wheat Board

The establishment of a marketing board for wheat was a milestone of singular importance in terms of state intervention in the nation's food economy. What happened with Canada's Prairie wheat economy was a matter that had been a major preoccupation of federal governments throughout much of the 1930s. With the push for a wheat board and the heavy lobbying over the minimum-price support for wheat, agricultural policy had become a significant, and divisive, political issue by the end of the decade.

In part, making a decision on wheat was important because the government of the day recognized that its policy would have implications for the state's role in the marketing of a range of other farm products as well. In a letter to Prime Minister King in March 1939, Agricultural Minister Gardiner attempted to convey the urgency of settling the matter of the Wheat marketing Board by arguing, "I am certain that further delay will make it more difficult to formulate policies suitable to other parts of Canada, in relation to the marketing of farm products" (quoted in Wilson, 1978:594). With an election looming, the perceived need to deal fairly with the Prairie producers and the communities they sustained occupied an extraordinary amount of the government's time and energy. Mackenzie King's diary entries for the period make this clear. In one entry he wrote: "I am far from sure that Gardiner is right in believing that we can get Western Canada to accept favourably legislation which will reduce price to be guaranteed farmers for wheat from 80 cents as it is now under the Wheat Board to 60 cents. I am afraid this legislation, if it carries will cost us many seats in Western Canada.... It will be, I fear, a sort of suicide to proceed with it" (quoted in Wilson, 1978:597). In the end, the issue of wheat marketing, and its implications for state intervention in other agricultural commodities, pitted cabinet minister against cabinet minister and posed a considerable dilemma for the government. Another one of King's diary entries revealed: "Discussion [in cabinet] largely upon budget matters and, in part, on agricultural programme, trying to find some means to overcome price fixing in connection with outlays by way of relief to Western farmers – an extremely difficult matter. The party is getting much divided – both in Council and in the house, and I fear also in the country over this business of fixing minimum prices for wheat, dairy products, fish, etc." (quoted in Wilson, 1978:599).

The Wheat Board affair also provides insights into the political process of formulating and implementing agricultural policies. By the 1930s, with the decline of the Progressive Party the agrarian producers had no substantial political presence in Ottawa, and this undoubtedly diminished their direct influence in securing policies reflecting their interests. Some years later, with the fortunes of the CCF on the rise across the country, the Liberals were no doubt feeling a threat on their left. Nevertheless, the CCF's strength was primarily at the provincial level, and its presence in Parliament was limited. Moreover, the CCF not only had the agrarian interests to promote, but was also increasingly concerned and oriented towards the problems of organized labour. To the influence of the CCF must be added the fact that the agrarian producers, especially in the West, still had their powerful co-operative organizations. Even when the wheat producers as a group were in serious distress with little in the way of personal resources to carry on a struggle, because of these organizations they could mount an effective lobbying effort at critical conjunctures.

Still, new scholarship makes it evident that the success of the Western farm community's program of securing wheat marketing legislation, given a limited presence on the national political scene, was dependent upon the shift in attitude of key sectors of the business community towards state intervention. This shift was itself largely a result of the breakdown of the capitalist market, so that even its chief benefactors – once normal times returned – had great apprehensions about the possibility of continuing with the status quo.

While changes in the strategic interests of the capitalist sector helped turn the tide in favour of state intervention, this sector also played a critical role in limiting the more far-reaching aspects of the proposed legislation. The capitalist sector moulded the final result to achieve a limited kind of state intervention that would only substitute for private enterprise when the private sector was manifestly unable to function in a minimally acceptable fashion.

It is also possible to overplay the role of business interests and their political representatives in determining the possibility and shape of agricultural policy development, if the case of the Wheat Board has anything to teach us. When the King government moved to retreat from its tentative steps towards state regulation of marketing, the agrarian producers' opposition to that retreat had sufficient potency to thwart a policy reversal. In addition, however much business as a whole had a general disinclination to state intervention, the capitalist sector was not monolithic. Some business groups on the Prairies actively supported the farmers in their efforts to block a retreat from intervention in the marketing of wheat.

Finally, during the war years the extraordinary conditions unleashed by the conflict largely negated whatever influence the capitalist grain interests might previously have had for limiting the state's role in wheat marketing. At the same time, the special wartime conditions that brought about the move to compulsory wheat marketing do not really confirm that the superior power of the agrarian interests had finally won the day.

Dairy Producers Achieve a Regulatory Regime

While agriculture in the Eastern provinces has not for a long time been as dependent on a single crop as Prairie agriculture has been, dairy production has been a mainstay of Eastern agriculture for over one hundred years. Indeed, the production of cheese was a substantial part of the 19th-century rural manufacturing sector, with almost 1,000 cheese factories located in the provinces of Ontario and Quebec by 1900 (Perkin, 1962:19). Cheese was also among Canada's top three export products before 1900, when Canada was supplying Britain with 60 per cent of its imported cheese (McCormick, 1968:16). Milk production was historically of first importance in all provinces east of Manitoba, and it is still the first or second most important agricultural commodity produced in every Eastern province (Ontario, 1986:Table 9). Therefore, the institutionalization of producer power through the development of a regulatory regime in the dairy industry was and is of no small consequence for the Canadian food economy.

By the time the Depression began, substantial structural changes in the dairy sector had already taken place. The change in the dairy industry of Ontario from a cheese-based industry to a fluid-milk industry was a result of many external factors. For one thing, the onset of World War I saw a decrease in cheese exports. Also, the war stimulated growth in other industries, and this increasing industrialization and its attendant urban development and population increase gave stimulus to the overall consumption of milk and dairy products. The growing consumption of fluid milk in particular created a year-round demand for milk producers geographically situated close enough to the major markets. This year-round demand was a key factor in bringing about the growing specialization of dairy farms and their shift away from the earlier mixed farming model. Prices for fluid milk were normally higher than those for cheese and other manufacturing products, and this plus the year-round demand gave farms supplying fluid milk to towns and cities an incentive to enlarge their herd size (see *Ontario Milk Producer*, 1930:118).[3] And while milk production on the farm was changing, on the industrial side cheese plants and fluid-milk dairies were being consolidated. Moreover, with the advent of the chain-store food retailers in the 1920s, a new and increasingly powerful player was being added to the milk industry.

In Ontario the Depression acted as a catalyst for the regulation of the dairy industry. Local market chaos and wide price fluctuations were predominant in many parts of the province. Farmers desperate to reduce their production surpluses began selling directly to consumers and/or to fluid-milk distributors at significantly reduced prices, while other farmers who usually supplied the cheese and butter plants attempted to get into the fluid-milk trade when prices for cheese and butter plummeted. This further depressed prices for whole milk (*Ontario Milk Producer*, April 1932:339, August 1932:43).

At the same time, dairy organizations were decrying the practices of the vendors of milk, and of the retail food chains especially. As the leading organ of the Ontario dairy farmers argued, "Reports come in of chain stores ... slaughtering milk prices, or declaring their intention to do so, thus precipitating a 'milk war' and a general cut in prices on the part of the regular distributors to meet the situation and hold their trade." This price cutting, it was noted, was reflected directly back to producers in the form of depressed prices (*Ontario Milk Producer*, March 1933:121).

The London, Ontario, market was particularly distressed. Farmers who in the 1920s were receiving about $2.12 per hundred pounds of milk were fortunate to get $1.00 per hundred pounds of milk after 1932. In fact, many farmers found they had surplus milk that they could only get rid of at seriously depressed prices, sometimes as low as fifty cents, or one-quarter of the pre-Depression price.

Even as late as 1939, the value of all milk and cream was only one-half of its peak in the 1920s, while butter and cream in the Montreal market were only 60 per cent of their 1929 prices. According to one student of the Canadian dairy sector, "By 1933 the industry in Ontario had become completely disorganized" (McCormick, 1968:21). For the dairy farmer, this disastrous situation revealed "the producer's powerlessness in dealing with monopolistic middlemen and processors" (Finkel, 1979:44), which had consolidated during the 1920s. These processing and distribution companies attempted to cover their losses from both the depressed economy and high tariffs on the international market by reducing the prices paid to the producers (ibid., 44).

In the case of the wheat economy, the elaboration of a state regulatory structure to overcome the manifest inadequacies and irrationalities of a market-dominated system owed much to the legacy of primary producer organizations – initially the grain growers' associations and later the producer-controlled co-operative enterprises. The salience of the producers' ability to organize themselves successfully was perhaps even more significant in the case of dairy, although the lack of scholarship in this area makes firm conclusions very difficult.

As many dairy farmers faced bankruptcy, producers who were previously complacent began to change their tune. As the normally staid *Ontario Milk Producer* argued by 1933, "What is needed today more than anything else is some influence or some movement that will increase the value of farm products, not lower them" (March 1933:121). In fact, it had already become evident to the more far-seeing fluid-milk producers that milk prices were no longer set locally, that prices set in the leading markets had an important influence in setting prices elsewhere. This, along with the consolidation of the processing sphere, got fluid-milk producers thinking about the need for a strong producer organization to safeguard their interests.[4] In Ontario, for example, the successful coalescing of dairy farmers producing for the fluid-milk trade in a producers league was significant for their eventual ability to protect their interests against an increasingly concentrated shipping, processing, and distribution industry.

An organization "born out of adversity and desperation" (McCormick, 1968:155), the Ontario Whole Milk Producers League set itself two main objectives: to correct the faults that affected the entire marketing structure of milk, and to standardize across the province vital practices such as the pasteurization of milk. The whole milk producers sought refuge in a regulatory regime not only because of the damaging price-cutting practices of some distributors and chain-store concerns, but also to prevent less responsible dairy farmers producing for the manufacturing trade from periodically jumping into the whole milk business and dumping their surplus milk. Also, with advances in transport and refrigeration technology, whole milk producers close to a dairy risked being displaced by more remote producers who were willing to provide the processor milk at a discount (see *Ontario Milk Producer,* January-February, 1933:111). It was with a view to ending cutthroat practices on all sides of the fence, therefore, that whole milk producers lobbied for some form of regulatory environment to provide what was coming to be known as orderly marketing.

A first success for the whole milk producers was the establishment of a milk marketing scheme by agreement of dairy producers and processor-distributors in the Toronto area. This scheme established a milk quota for each dairy farmer supplying milk to Toronto dairies, based on the quantity of the producers' milk, the regularity of supply over the previous year, and the length of time the producer had shipped to the dairy.[5] The dairies, for their part, agreed to pay producers a fluid-milk price for milk supplied within the quota, and a somewhat lower price for milk that went for cream. Surplus milk, that is milk above quota, received another lower price. The producer had to agree to ship to one dairy only, and any surplus milk had to go to the same dairy. Breaking this last rule would bar the producer from shipping to Toronto dairies.

With this scheme producers and processors alike hoped to stabilize the milk trade in the Toronto region and put an end to the chaos that was then the norm because of producers regularly dumping their surplus milk at depressed prices and the distributors' subsequent underselling of the product (see *Ontario Milk Producer*, June 1933). At the same time the Producers League began an education campaign aimed at getting whole milk dairy farmers to utilize any surplus milk on the farm, as feed for chickens or livestock, rather than bring it to market and further destabilize an already depressed price structure.

The Milk Act and the Milk Control Board

While the Toronto market was being brought under control, the same could not be said for all of the forty or so distinguishable markets for milk in the province. It was in large part to bring some order to the chaos reigning in a number of these local markets that the whole milk producers association lobbied the provincial government hard for legislation to stabilize the industry. For their part, the processor-distributors also appeared to want some form of government intervention because the dumping of surplus milk by cheese producers and the cut-rate selling of the surplus milk by "fly-by-night" distributors were endangering their own profitability.[6] The producers, through the pages of the *Ontario Milk Producer*, argued that certain injurious practices were damaging the industry and urgently needed to be addressed by government policy. These practices included: (i) distributors offering privately, and producers accepting privately, a lower price than that bargained for by the associations; (ii) distributors declaring as "surplus" milk, and paying the lower price accorded to it, an unduly large proportion of milk shipped; (iii) distributors failing to pay for milk shipped; (iv) distributors inducing or compelling shippers, as a condition of gaining or retaining their market, to purchase stock in a dairy; and (v) distributors' culling of retail prices, either directly by selling at a price lower than the agreed price, or indirectly by offering rebates, discounts, or premiums (Hennessey, 1965:25-26; *Ontario Milk Producer*, June 1935:11).

The Milk Control Act of 1934, amended a year later, was a reflection of this pressure, without going as far as the producers would have liked. One of the advances, from the dairy farmers' point of view, was that the Act required all distributors of milk for human consumption (processors and retailers) to be licensed and, furthermore, to be bonded. The bonding requirement was to ensure that farmers got paid for the milk they shipped, which in the past had not always happened. The licensing, in theory at least, gave the government some leverage over distributors in the process of coming to a mutual agreement with the producers in each local market concerning an acceptable price and ensuring that all distributors abided by the agreed-upon price. Where producers and distributors could not agree on price, the Milk Control Board established by the Act had powers to act as arbitrator.

The Milk Control Board did not have powers to fix prices for milk, but it did have the authority to discipline, through the threat of, or the actual, revoking of distributors' licenses if their business practices served to undercut and disrupt locally negotiated prices.[7] The actual price to be paid producers in each local market was still, nevertheless, one that had to be negotiated by the local producers and processors themselves. While the Board was expected to use its "good offices" to facilitate an amicable settlement, there were cases where a price satisfactory to producers was not to be had. Here it was not clear that the Board had the authority to solve the producers' problem. There would still seem to have been considerable leeway for processors to keep prices low, at least some of the time in some local markets.[8]

In the Toronto market, prices by 1934 had advanced to $2.10 from $1.45 per hundredweight of milk between 1932 and 1934, suggesting that the successful establishment of a quota agreement, together with the influence of new legislation to regulate the marketing of milk, had a positive impact on price (Ontario Milk Producer, January 1936:154).

In the years that followed, the regulatory regime relating to the dairy sector became somewhat more complex, and modifications were introduced to reflect changing economic circumstances. The dairy producers were unhappy with agricultural prices set by government fiat during the Second World War period and were able to get price supports – via the Agricultural Prices Support Act of 1944 – to prevent a collapse of farm prices like the one that had occurred after the First World War. Price supports were further institutionalized in the late 1950s with legislation that had the objective of maintaining a "fair relationship between prices received by farmers and the costs of the goods and services they buy, thus to provide farmers with a fair share of national income" (quoted in McCormick, 1968:58). The Agricultural Stabilization Board guaranteed the prices of butter and cheese, among other key commodities, at a level that was 80 per cent of the average price realized over the preceding ten years.

By 1957 producers had managed to secure a formula pricing policy in the new Milk Industry Act. This policy established the principle that milk prices should bear some relation to farmers' actual costs of production. In reality, there were still price differentials around the basic formula price in each of the local milk markets (ibid., 79-80). This situation was to change dramatically with the move in the late 1960s to establish a pooling system for milk that encompassed the entire province.

The Move to Supply Management

Curiously, it was under the Tory government of John Robarts that Ontario made a concerted attempt to subordinate milk's multitude of private market arrangements to an all-encompassing system of supply management – though we should recall that the first major steps towards state intervention in the agricultural sphere at the national level were under the auspices of the Conservative Bennett government of the 1930s. The Bennett government represented an old-style conservativism that saw the state as playing an important role in establishing social peace, as opposed to the "neo-conservative" orientation of the Progressive Conservative party today, with its anti-statist, pro-market position more closely aligned to the classical liberal views associated with Adam Smith.

Still, by the mid-1960s there were some 179 separate marketing agreements governing the sale of fluid milk in the province, providing for eighty-three different prices. Producers in many areas still feared losing their markets if they didn't cater to the wishes of certain dairies and even to the milk transporters, who enjoyed considerable influence because of their role, together with the dairies, in determining who got the fresh milk quota and who did not. Some lucky producers were getting special deals from dairy processors; others had to give significant kickbacks under the table to dairies to maintain their market. Quota abuses abounded in the fluid-milk market (Biggs, 1990:58, 61, 63-64). One of the few accounts available describing the milk industry during this period, written by a former Deputy Minister of Agriculture, cites the experience of one dairy farmer: "I had quota, and half of it started to be returned coloured, indicating unacceptable quality. There was no way. It was all the same milk. The dairy was just colouring the milk they didn't need and rejecting it. They were buying non-quota milk from around (the town) at about half the agreement price. They were sending the quota milk they should have paid for back to the farmers" (Biggs, 1990:59).

In the mid-1960s the Minister of Agriculture, Walter Stewart, guided by the Hennessey Inquiry into the milk industry, moved decisively to establish an alternative to the prevailing private milk marketing system. Legislation was enacted with the establishment of the Ontario Milk Marketing Board (OMMB) to end the abuses endemic in the existing arrangements and to provide for equitable treatment of the dairy farmers. The first chairperson of the Board, George McLaughlin, made the intention of the legislation clear early on when he stated, "We're saying the milk producers in Ontario are going to get a stronger collective countervailing power in the market place. We're going to get improved prices for milk" (quoted in Biggs, 1990:60).

The Board's intention was to bring in a pooling system for milk. The Board would supply the dairy processors with the milk, the money would all be put into one pot, and producers would all be paid the same price for their milk. It was not long before the vested interests in the dairy industry moved to try to block the plan, but the Board moved ahead in any case. The pooling idea for milk was first tried out on a pilot basis in Northern Ontario. Its success there encouraged the Board to forge on with a Southern Ontario pool in 1968. Crucial to the success of the Board's ambitions was the setting up of local milk committees at the county level to act as a vital liaison between the Board and the producers.

In March 1968 the pool quota order detailing the individual quotas of more than 8,000 fluid producers was published. These quotas were based on a nine-month period during 1965-66 indicating producers' historical production. Quotas were negotiable, and all producers within the pool were to be paid on the same basis. At the same time the Board simplified the 170-odd distribution areas into ten new ones and took on the task of rationalizing the trucking of milk, which had been subject to tremendous duplication of routes and overall inefficiency. This step alone saved producers hundreds of thousands of dollars. In addition, a protection fund created by provincial legislation for milk and cream producers protected the dairy farmers against plant failures (Biggs, 1990:101-2).

The next step was to bring industrial milk under the Board's control. If there had been ongoing problems with fresh milk marketing, the problems were even more apparent with industrial milk. With fresh milk, for many years before the Board's creation producers had developed some form of quota system, albeit at the local level, to limit the production of surpluses. This had not been the case with the 14,000 or so farmers producing milk for "industrial" purposes, such as the manufacture of cheese. One writer well versed in the realities of the industry at that time notes: "Traditionally there had been uncoordinated processor demands encouraging industrial milk production and the resultant unmanageable surpluses of milk products. There had been a continuing problem not only for producer prices but also for the federal government, which had attempted to store the surplus products and dispose of them at significant losses on the international markets. In some cases they gave them away" (Biggs, 1990:104).

In addition to problems created by chronic overproduction, there were reports of rampant inequities in the dealings between producers and processors, with some producers getting preferential treatment in one form or another. There was also little control over such matters as butterfat tests in the plants, tests that governed what prices producers would receive (ibid., 107).

The OMMB intended to rationalize the industrial milk industry by bringing in a market-sharing quota system (MSQ) for industrial milk, which was developed in co-operation with the Canadian Dairy Commission and leaders from dairy producer organizations in Quebec. The MSQ was essential if the chronic problems of oversupply and low prices for farmers were to be avoided. With this arrangement, not only could surpluses be controlled, but also each farm would share equitably in the market available for industrial milk. Producers with an industrial milk quota paid a small levy (twenty-six cents per hundredweight) on milk produced within quota to cover the costs of industrial milk exports. An overquota levy, essentially a penalty for producing a surplus, was charged for milk produced outside the quota, at a rate of $2.40 per hundredweight.

Besides the MSQ it was felt that supply management would only work in the industrial subsector with a plant-supply quota policy. This meant that all processors producing cheddar cheese, butter, powder, and condensed milk were allocated a quota based on the amount of milk handled over the previous year. Together with the plant quota went a new classification system governing the price that producers could receive for milk going for different types of industrial milk products. Under this system, the guiding principle for the Board was to allocate milk first to those plants producing products that realized the highest returns in the marketplace. The remaining milk was allocated to plants according to the plant-supply quota policy. What this system meant in reality was that firms producing products such as cheese, which fetched a low price in the market per unit of raw milk utilized, did not always receive the milk they wanted. Indeed, industrial milk processors lost the prerogative they had formerly had to encourage the production of the supply they wanted and to take whatever part of this supply they felt they could utilize at any one time. With the Board's supply management system, it was the Board and the plant-supply quota system that regulated the production of each plant, and no longer the vagaries of the open market. Not surprisingly, some private processors were not especially supportive of the scheme. Some did what they could to block it, and so did some of the more privileged dairy farmers.

Resistance and the Role of the Provincial State

A few influential dairy processors saw the new scheme as taking away their prerogative to deal with producers as they saw fit. The businessmen behind the expanding Beckers Company retail milk chain stores were especially opposed to the scheme. They were aggressively marketing milk in retail outlets directly to customers, a practice some saw as undermining the long-established home delivery system. The few large dairy farmers with lucrative contracts with Beckers also feared a loss of their privileges and organized to block implementation of any regulations, such as the milk quota, that would stand in the way of an expansion of their operations (Biggs, 1990:91-92). So too did farmers specializing in the Channel Island breeds – Jersey and Guernsey cows – who were getting preferential prices for their milk. Under a one-price-for-every-producer system they feared they would lose out, and their organization fought the scheme bitterly, all the way to the Supreme Court of Canada.

Given the disorganization of the milk industry as a whole, and especially the splits that characterized even the producers as a group, it is surprising that a regulatory regime as all-encompassing as that represented by the OMMB could be put in place at all. This major example of the institutionalization of producer power illustrates particularly well the creative and almost independent role the state can play, at certain times, in providing stability in capitalist society. In the jargon of political sociology, this "relative autonomy" of the state – as represented here by the OMMB – can be a crucial factor in our understanding of how certain events can transpire.

The state's unique role in this situation was predicated on the existence of a certain set of conditions. In the first place, there was general confusion in the industry, with little prospect of any one sector imposing its will on the others to bring about stability. The dairy producers were seriously divided among themselves, unable to provide a united front to push their interests coherently, because these interests were not one. The income of the fluid-milk producers, for example, depended largely on the success or failure of their distributors. Some dairy farmers, especially large ones, benefited from special deals with processors, while Channel Island-breed producers had worked to establish a preferential situation of their own with the dairies. Moreover, the conditions affecting producers in the far western Rainy River district were not the same as those of northeastern Ontario, let alone those of the Toronto area or southwestern Ontario. Even more significant were the different interests of fluid-milk shippers and the producers whose milk went for cheese production and other industrial uses. Prices negotiated for fluid milk were historically

much higher than those for "industrial" milk. This, plus the fact that entry to
the fluid-milk trade was highly restricted, created considerable envy and resentment on the part of non-fluid-milk producers (Hennessey, 1965:34). It
was difficult for any one overarching producers' organization to effectively
represent the interests of such a diversity of producers.

In fact, four separate organizations continued to exist representing different types of dairy producers, and these organizations had shown little ability
to work together to protect the interests of dairy farmers as a whole. This was
despite the growth of multiproduct processing plants, which increasingly
made it more and more urgent that industrial and fluid-milk producers come
together in one organization to effectively protect their interests vis-à-vis the
large processors. By the late 1950s the executives of the four dairy producer
organizations had formed the Ontario Milk Producers Co-ordinating Board.
Their objective was to get all four organizations working together to develop a
scheme to put the milk industry on a more rational and equitable basis. However, despite a concerted effort, all four organizations could not agree on the way
ahead, and the Board turned to the provincial government for direction.[9]

The side of the processors seems to reveal a certain lack of common cause.
The process of consolidation was creating a few large processors, whose interests were not necessarily synonymous with those of the smaller local dairies
that continued to hang on. The new processors with retail outlet operations,
pioneered by the Beckers Company, had different interests than the standard
dairies, who were preoccupied with the prospect of this new development
making serious inroads on the long-established home delivery system of marketing milk. Cheese processors faced different problems than did the fluid-
milk processors. Also, this subsector was itself increasingly divided between
the multitude of cheese factories, many of them very small operations, and a
few giants engaged in the "cutting and wrapping" business. The function of
these larger companies was primarily to package cheese purchased from the
smaller factories and distribute the product to supermarket chain stores under
well-advertised brand names that the "cutters and wrappers" typically controlled themselves. The U.S. multinational Kraft was and continues to be a
dominant player in the "cutter and wrapper" end of the cheese business.

Given this multitude of competing interests between producers and processors, and within these groups as well, it is understandable that only an *external force* was ever likely to be able to affect a fundamental reorganization of the
industry that would in turn provide a lasting stability. It was left to the provincial state to provide the unity of purpose that would achieve this goal.

The question remains as to why the state moved to establish an all-encompassing supply management system when it did. The problems of the milk industry were, after all, not exactly new ones. They had existed in one form or another for decades. What, then, prompted the provincial state to intervene so decisively at this particular time? An answer to this question is difficult to find, largely because research on the subject is still inadequate. But a few elements of the conjuncture that produced the OMMB would appear to have been crucial.

An adequate explanation would have to look at both personal and broader structural factors. The role of the new agricultural minister – Walter Stewart – was notable, for the success of a scheme like the OMMB required a minister who had the vision to see the value of a pooling scheme for the industry and the wider public as well. Furthermore, it required a minister with the determination to make the scheme work. This meant protecting the new Board from the inevitable enemies that such a far-reaching plan brings forth. Stewart clearly had these qualities.

It must also have been important that a major element of the processor interests – notably the large multiproduct processors – were not adamantly opposed to the scheme; that they were indeed willing to go along with it as long as they were sure that the government's involvement would prevent the producers from using their powers unilaterally to reorganize the industry. The big processors were also keen to bring some order to the chaotic situation in the milk transport business and to achieve improved security of supply. The new scheme offered to solve these problems. This acquiescence of most of the major players in the processing end was most likely a necessary ingredient for the Conservative government of the day to take the lead it did.[10]

The United and the Divided

The struggle by agricultural producers to institutionalize their power has proceeded at a different pace in each sector of agriculture, and in the end the producers of certain commodities, such as beef, have resisted the general trend. Even producers of a single commodity have galvanized different instruments to achieve their ends, as was the case with wheat producers, who have utilized both co-operative organizations and a federal regulatory board to protect their interests. Although a system involving a provincial milk board with legislated powers developed in Ontario, elsewhere, notably in Nova Scotia, Quebec, the Prairie provinces, and British Columbia, producers established co-operative dairy-processing plants to help solve their problems. For some time producer co-operatives have had the dominant market share for fluid milk in most of these provinces. In the post-World War II period, producers of the rapidly growing broiler chicken sector and egg producers also fought for and won a system of supply management to stabilize their industries. Producers in other sectors, such as those in Ontario who produce vegetables and fruit for the processing industry, organized beginning in the 1940s to secure their interests through marketing boards with provincially legislated powers; this has given the farmers leverage in negotiating prices and contract conditions with food processors.

The experiences of wheat and dairy producers in trying to protect their interests suggest some notable conclusions. The relative unity of the wheat growers was imperative in securing for them, before most of the other commodity groups, government-supported mechanisms that would help secure their objective of orderly marketing. The more fractious producer groups, such as the dairy farmers, had to wait much longer to achieve a degree of state intervention to secure their interests, if indeed they were successful at all. Moreover, it would seem that where producers are divided, their chances to successfully achieve a degree of protection from the capitalist market has been dependent upon the coming together of a favourable conjuncture of events, a conjuncture that included at least the acquiescence of capitalists in the processing sector, and furthermore the willingness of the state – at either or both the federal and provincial levels – to intervene to coerce a degree of unity among the various actors involved.

However, in the cases of the Wheat Board and the Ontario dairy producers, it is clear that the full objectives of producers were not achieved. In both cases producers were fighting for concessions from governments that were beholden to the big business community and therefore sympathetic to the private sector's lobbying efforts to defeat the more far-reaching proposals of the producers. This fact points back to the significance of the producers' failure, at the federal level and in most provincial jurisdictions, to mount a political vehicle that would advance and protect their cause. Without such a vehicle, they left themselves critically vulnerable to the expanding agribusiness complex that has come to dominate the food system in recent decades.

While the combination of producers' co-operatives, supply management regulations, and marketing boards has helped shore up the power of primary producers in the last half of the twentieth century, the privately controlled organizations that constitute much of our food system have played a dominant role. Their concentration and growing market power formed a major impetus behind producers' efforts to protect themselves behind co-operatives and regulatory regimes to establish "orderly marketing." It is to the rapidly developing sectors of the food system controlled by private capital that we now turn our attention.

PART II

THE AGRIBUSINESS COMPLEX: A NEW LOCUS OF POWER

••••••••

THE TRANSFORMATION OF FARMING AFTER MID-CENTURY

Some time around 1950 in Canada, farming as a unifying activity for a substantial proportion of the population entered a period of rapid change. This change was impelled by technological and economic forces, and it has had profound social and political consequences for the farm community and beyond. As Don Mitchell argues in his now classic study of the Canadian food system, *The Politics of Food*, "Until the end of the Second World War farmers were able to identify certain common enemies and their concern centred primarily around issues of marketing their products. They had solid social communities from which to organize and they shaped strong and united farm organizations and political groups which won important concessions over the years. After 1950 all of that changed" (1975:14).

The question of *why* things have changed so much is a big one, and I will devote much of the rest of the book to exploring some of the most important factors at work. As for the question of *how* things have changed for farming – this one is perhaps a bit easier to answer, and it is especially important to consider if we are to appreciate the magnitude of change and the seriousness of the problems that follow from that change.

As Mitchell notes, it was only after the demise of horsepower around mid-century and the lifting of physical limitations on how much land could be farmed that the costs of land, machinery, and farm production began to increase uncontrollably for farmers and became a major problem. A major technological

revolution began to transform farming, especially with the introduction of mobile forms of power, particularly the tractor (see Winson, 1985).

Many of the problems faced by farmers in the postwar period can be traced
to what has been called the "cost-price squeeze," described by Mitchell as "a
process in which the rate of increase in the combined costs of producing a
commodity were rising faster than the gross return received by the farmer for
the commodity when he sold it" (Mitchell, 1975:18). Between 1981 and 1989,
for instance, farm input costs rose by 16.5 per cent, while farm product prices
by 1990 had actually declined from their 1981 levels. For such major Canadian
agricultural commodities as wheat, the price decline over this period had been
much greater, at about 22 per cent (Ferguson, 1991:7). The cost-price
squeeze, then, has posed a major dilemma for Canadian farmers over the last
several decades, and continues to do so today.

As a result the net income of many farm families has been substantially
eroded, and strategies to resist this trend have altered the makeup of the farm
economy. For the most part, the forces pushing net farm income down were
met by attempts to increase the volume of production on the farm. With the
"tractorization" of agriculture and a dramatic increase in the use of chemical
sprays it became possible, at least for some, to work much more farm land
without raising the input of increasingly expensive farm labour. The incorpo-
ration of ever greater volumes of chemical fertilizers and other inputs, such as
hybrid seed varieties, helped boost yields per acre. These strategies allowed
farmers to advance their productivity by an average of 6 per cent per year between 1945 and 1970 alone.

The growing pressure on farm incomes pushed producers to follow other
strategies as well, such as formalizing relationships with food processors to
supply raw produce under some form of contract arrangement. Whatever
strategies were pursued, however, it is clear that not all farmers were able to
adopt them, and as a result there has been a definite process of *social differentiation* within the farm community. Some farmers had the necessary land base
or were able to borrow to purchase land and expand the volume of production
on the basis of increasing their debt load. Others adopted a more conservative
strategy of avoiding debt, conserving farm machinery, and attempting to get
by with less. Many found these options unworkable and decided to leave
farming altogether (see Hedley, 1979; Mitchell, 1975).

In general, as Mitchell noted some years ago, "The conditions surrounding the cost-price squeeze were controlled and directed in such a way that
farmers ultimately could only be the losers" (1975:20). The number of farms

peaked in Canada in 1941 at 732,832 and then dropped by some 60 per cent to 293,089 in 1986 (Canada, 1987:2). The farm population fell by an even greater proportion: 67 per cent over this period (Basran and Hay, 1988b:22). In some regions, notably the Maritime provinces, the decline was even more precipitous, with the number of farms declining by some 85 per cent since 1941. In that region the magnitude of the decline suggests the virtual failure of a regional agriculture (see Winson, 1985).

While farms have been disappearing at a rapid rate, the land they once controlled has for the most part been incorporated into the remaining farms. Average farm size increased by 66 per cent from the mid-1950s to the mid-1970s (Canada, 1987:4). This process has been part of the concentration of production within a relatively small, privileged stratum of Canadian farms, which now control a disproportionate amount of land and rural resources. By 1981 about 25 per cent of farms in this country accounted for 74 per cent of gross farm sales (Bollman and Smith, 1988). Within this stratum of the largest commercial farms there is an even smaller group of "super-farms" that has emerged in recent years, and which accounts for gross annual sales far out of proportion to their actual numbers. By 1981 the top 1 per cent of census farms controlled 19 per cent of aggregate gross farm sales (Stirling and Conway, 1988:76). These farm operators have often spawned a number of agribusiness enterprises beyond the farming operation itself. Bob Stirling and John Conway provide an interesting example of this type of emergent super-farm:

> One example is the Lakeside Feeder "empire" of Brooks, Alberta. It was created when three budding entrepreneurs joined forces in the mid 1960's. The firm includes 2,000 acres of irrigated land, an 18,000 head feeder operation, a feed mill, a 450 sow hog operation and a chicken business that produces 600,000 fryers annually. In 1974, Japanese capital (Mitsubishi Canada) took a 20 percent investment in the company, allowing the partners to build a 300 head a day slaughter plant for beef....
> The company, composed of seven different enterprises, also sells fertilizer and herbicides. Three hundred people are employed. (1988:77)

The super-farms could be expected to share the concerns and ideological outlook of the people who run large agribusiness enterprises. This elite of farm operators is more likely to see capitalist agribusiness firms on the output side – processors, exporters, dealers, and exchange brokers – as their natural allies than the farming community as such. While this type of massive accumulation is taking place among a privileged few at one end of the farm spectrum, a very much larger number of small farm operators have had to seek off-farm work in ever greater numbers in order to stay on the land. This reached the point in the 1980s where 40 per cent of farms reported off-farm work in Canada as a whole, with the proportion reaching over 50 per cent in some provinces (Basran and Hay, 1988b:7).

Inevitably, then, the acceleration of the process of social differentiation within the Canadian farm community has an impact on the solidarity of this class and, by extension, on the effectiveness of producers when it comes to protecting their way of life in a politically coherent manner. Stirling and Conway sum up the significance of this process for Prairie agriculturalists:

> The trends in the prairie farm class – the growth of both small and large farms, and the appearance of the "disappearing middle" phenomenon – suggest that the recent crisis in prairie agrarian politics is no short term affair. It is, in fact, the beginning of an historic re-orientation of farm politics from a politics based on the assumptions of homogeneity and similarity of interest to a politics based on the clear class cleavages which have irreversibly emerged among prairie farmers. Whether this new reality will be addressed by a new leadership and a new organizational thrust uniting "small" and "middle" farmers on a clear class program will have significant implications for future politics on the prairies. (1988:82)

Again, this discussion relates to the question of *how* farming as an activity has been influenced and altered by external forces in recent decades – a matter on which interested readers will be able to find a substantial literature dealing with developments in both Canada and the United States. As to *why* such changes have taken place, there is much less written, especially in the Canadian context. The following chapters explore the development of the Canadian agribusiness complex, wherein lies at least some of the answers as to why farming has been so transformed. Moreover, the dilemmas posed by the contemporary development of agribusiness have a relevance well beyond the food system, as we shall see.

Farming, Banks, and the Cost-Price Squeeze

Susan, an Ontario farm woman, and her husband Peter raised pigs on a family farm owned jointly with Peter's parents. Together with Peter's brother Carl they expanded operations to a second farm. This is her testimony:

Peter and his brother Carl started with a fifty-sow herd. They started slowly buying sows and then they decided they needed to renovate the barn so they went out and got a bank loan. They got a $40,000 loan. That was the beginning of the end, because that was the beginning of the debt load.

I was working in town at a drug store at the time, and they asked me if I would quit and take care of the farrowing room. And I thought, yeah, that sounds great. It was 1980. I started working in the barn and I was reponsible for the sows when they farrowed, for teeth, tails, iron shots, the vaccination program for the whole herd, castrating, cleaning pens, helping move pigs with Peter.

Back in the eighties, the money seemed so easy to get, and you had all these big dreams like other people do – to enlarge, get better equipment, or just to improve. So you went to the bank and they would lend you money. Peter and Carl went to the Farm Credit and said, we want to buy another farm, we want to increase the herd.

We had the mortgage on the family farm and an operating loan. We were farming for thirteen years ... and one year we made a profit – it was small, maybe $3,000 – and that was the year we got involved with some government programs. The program was called OFAB option C, the government would guarantee your operating loan from the bank for three years. Our operating loan at that time was $95,000. We went into the program with the thought that the government would guarantee this $95,000 and the bank would continue giving us money and continue to carry our debt load.

Everything went along really well. When we went for refinancing at the end of those three years we had $20,000 in the bank and we took it to our bank manager, who we were on really good terms with ... and we said, "We'll give you this money and if we show a profit this year, will you finance us for our next year?" They said, "Yes," and we shook hands on the deal and we left thinking that the financing would be in place.

In January the loan ran out and they told us they were calling in the loan, just like that. We had never missed a mortgage payment or an interest payment, and at that time we were paying $1,400 a month in interest alone – and that was one payment we knew we could never miss. As far as we were concerned, we were in good credit standing and they just came in and said, no, we are not going to refinance you and we want the money back.... They got all their money back and didn't have to take a risk at all.

In December 1985 wiener prices were $1.08 a pound, or $2.38 a kilogram. In November 1992, which was the last wiener shipment we made, we got ninety cents a pound, or $2.08 a kilogram. In 1985 a ton of soybeans was $299, in 1992 it was $318. In 1985, a bag of "super swine," which is the mineral supplement you mix with soybean and corn to feed the pigs, was $13.50 a bag. In 1992 it was $15.50 a bag.... We were spending, with a fifty-sow herd, $300 every two weeks on premix.... I remember one time when we were having health problems in the barn, our vet fees were $2,500. And our hydro bill in 1985 – and that was running two barns – was $984 for three months. In 1992, with one barn empty, it was $1,385.

Pig farming, beef, cash cropping – because there is no quota for them there was absolutely no protection – the price was the price. The price has never been good. It has mostly been below the cost of production.

So there we were, they were calling in our loan. And yes we were in financial trouble, we were really highly leveraged. So we called the Farm-Debt Review Board. The FDRB wanted us to sell our pigs. They wanted us to sell the second farm, and we did that, so that paid off the one mortgage we had on that particular farm. They came and took all of our equipment: the combine, the tractor, and even an old garden tractor that we had in the barn that didn't even run. They wanted to take the feed mill out of the barn, but it was cemented to the floor, so they couldn't take that.

It was phenomenal, what went on. They wanted us to sell our hogs but we didn't want to because we wanted to keep farming. So we sold them to the owner of the feed store and we bought them back from him on a monthly payment, which we are still paying.

The FDRB is supposed to be co-ordinating efforts to save you from bankruptcy, but looking back at it, they didn't do a hell of a lot. We thought, what are we going to do? We will have to buy corn for most of the year, and we just can't do that any more.... Finally we just decided to chuck it all.

It was really hard for Peter to make that decision. He went through a real grieving process of getting over it. We had no idea of what we were going to do with our lives, where we were going to live. We'll have to get a small apartment and just start saving, but what about Peter's parents? We could put them in a seniors' home, I suppose. It's the best thing for them and it is is cheap rent.

It was awful. When I look back at the things we were doing, we were drinking a lot just to get over it. We were fighting all the time about money. You couldn't buy anything, you couldn't do anything, it was awful. It was the most intense part of my life ... it lasted for about a year and a half. At night when we would go to bed, Peter would be just beside himself. He had always had the burden of his parents because they put money into the farm. There was a lot of responsibility there for him.

I think it is really sad what happened to us. We were good farmers, and if we had been making a good buck for our product we wouldn't be in the financial straits we were in. Yes, it would have always been tight because we were always really highly leveraged, but we would have paid off our debt eventually, at least worked the debt down and lived a good lifestyle. But now we are not farming any more ... it is not the Canadian dream.

Source: Interview, March 1993.

CHAPTER FIVE

CORPORATE CONCENTRATION IN THE EARLY 20TH CENTURY

• • • • • • • •

*An economic collapse may, like rain, fall both on the just
and the unjust, yet our evidence proves that there are
some groups in our economic system able to escape the full
force of the crash. Often the way of escape is at the expense
of the other groups less powerful and, therefore,
less fortunate.*

Royal Commission on Price Spreads, 1937

In the early decades of this century the struggle of the farming population for recognition of its important place in Canadian society and the agitation of its organizations for economic, social, and political reforms were high on the national agenda. Indeed, the impetus for this struggle came at least in part from the rapid structural changes of Canadian society, the frenzied expansion of urban manufacturing based on mass-production technologies under the umbrella of the National Tariff, and the unprecedented growth of cities – developments that farm people took to be largely at their own expense. There were also developments that more directly changed the fortunes of farm families.

In the decade before the Great Depression, the Canadian food economy had in an important sense been transformed, especially those sections of the food economy situated beyond agriculture *per se* and for the most part organized by capitalist enterprise, and large capitalist enterprises at that: the sections concerned with the transformation of food, supplying farmers with inputs, and organizing the distribution and sale of food to the consuming public. These food sectors had an increasing influence on the marketplace and on determining the nature of the goods produced by 20th-century agriculture. And even by the third decade of the century, a small number of firms in these sectors had garnered a large proportion of market power. In the end, the structure that was established in this early period would necessarily shape the post-Second World War food economy.

The Historical Phases of the Food Economy

Even in the most economically developed countries, the rise of agro-industrial activities to an economically dominant role – when compared to more traditional farming – is itself a fairly recent event. It is a phenomenon that has occurred largely since the end of the Second World War. Nevertheless, the roots of this development were laid in an earlier period, in what Louis Malassis in his pioneering work on the evolution and structure of the food economy has termed the third phase or stage of its development.[1]

Malassis argues that the first stage is the pre-agricultural food economy (*l'économie alimentaire pre-agricole*), in which the primary methods of obtaining food were gathering (berries and other fruit, nuts) and hunting and fishing. The second stage – the agricultural and domestic food economy (*l'économie alimentaire agricole et domestique*) – is associated with the successful domestication of plants and animals. The greater part of activities within the food chain took place within the consumption unit, which was typically the unit of agricultural production as well. This unit would be more commonly known as the peasant household. Malassis dates this stage from neolithic times and as having existed in one form or another until its transformation beginning in the 19th century.

The third stage he identifies is what he termed the commercialized and diversified agricultural food economy (*l'économie alimentaire agricole commercialisée et diversifiée*). The 19th century witnessed the rapid development of this stage as peasants produced for the market more and more while the subsistence economy of small holders went into decline. Towards the end of the 19th century there were major advances in transportation technology (railroads, iron ships replacing wooden sailing vessels). This spurred a rapid internationalization of commerce, as the cost of the long-distance transport of agricultural produce declined dramatically. The opening up and settlement (typically through the use of force against indigenous peoples) of new lands – the North American Prairies, the Argentine Pampas, Australia – together with advances in the field of transportation established a historically unprecedented global commerce in basic foodstuffs. This global commerce in products such as wheat and chilled beef had the effect of restructuring diets for much of the European population. At the same time it placed strong economic pressures on different forms of agricultural production – principally large-scale capitalist farming and peasant farming – that proved to be less efficient than independent family operations in the Americas (see Friedmann, 1978a). Under this pressure these forms of production – of fairly recent origin in the case of capitalist agriculture – began to disappear.

In this stage the purchasing power of the mass of the population in the most developed countries increased sufficiently to allow for a shift from relatively inexpensive calories (cereals, root crops) to more costly calories (meat and dairy products). The social and technical changes that made possible the industrial revolution also made possible the "industrialization" of food. The first forms of agro-industry took shape.[2] However, rapid developments in transportation technology and the spread of assembly-line techniques pioneered in the fledgling automobile industry were soon to make possible the development of larger factories with an extended market reach. The ravages of price cutting and overproduction spurred capitalists in this sphere of the economy to circumvent competition through mergers that created much larger, multiplant companies by the early decades of the new century.

In Canada, Malassis' first stage characterizes the economy of some of the indigenous peoples, while his second stage describes the food economies of some indigenous tribes that had begun the practice of sedentary agriculture, as well as, up until the 19th century, the seigneurial rural economy of what is now the province of Quebec. In the rest of Canada, the relatively late European settlement and the modern form of landholdings that characterized it in comparison to Quebec meant that the food economy took on the basic characteristics of his third stage. By the 1930s this stage had reached a high degree of maturation in Canada.

The Early Concentration of Capital in Canadian Agribusiness

A main justification for organizing economic and social arrangements in society along capitalist lines, with private ownership of the means of production, centres on the role of a competitive marketplace. Where sellers and buyers of goods compete on a free and equal basis, it is argued, productive resources are allocated in the most efficient manner and low prices to consumers are the result. It is also argued that a competitive market economy is the best way to maximize income in society. These happy results, which received their classic statement in Adam Smith's *The Wealth of Nations*, published in 1776, were predicated on the existence of a state of *free competition*. This state may be said to exist when the producing firms in the industry are sufficiently small that they can have no significant effect on the market by withholding supply; that is, in return for their products they must accept the prevailing market price. Moreover, no buyer has enough control to have a hold on sellers in the market. Finally, the industry is characterized by a freedom of entry and exit; that is, existing firms do not have the power to effectively bar the entrance of new firms to the field.

There is evidence to demonstrate that the idea of this competitive market did not accurately apply to the Canadian economy by the 1930s, if indeed it ever did. In the words of the 1937 Royal Commission on Price Spreads, which took up the question of the socio-economic consequences of corporate concentration, by the 1930s Canada was "a transitional economy in which simple [free] competition still prevailed in some parts, monopoly had succeeded it in others, and monopolistic or imperfect competition characterized the rest" (Canada, 1937:6). Indeed, four or fewer firms controlled the great majority of output in many of the main sectors of industrial manufacturing in Canada. The corporate merger activity that produced this situation had in fact proceeded at a frenzied pace in the years just preceding the "Big Crash" of 1929 (see Veltmeyer, 1987:28-29). As a result of this merger activity, by 1929 plants with a total of $1 million in output, just 3.2 per cent of the total number of plants in the country, accounted for 62 per cent of total output in the economy.[3]

The Canning Industry Consolidates

The growing food-transformation industry was not an exception to this trend to concentration of production, and indeed in this sector the impact of corporate merger activity was especially significant in view of the continuing prominence of agriculture for Canada as a whole. The extent of early agribusiness consolidation is starkly illustrated by the history of the fruit and vegetable canning industry. Quite early in the century many farmers in Ontario, the Maritimes, and British Columbia were dependent upon the canning industry as a market for their produce. Even by the turn of the century, canning capitalists in Ontario seeking to avoid the ravages of competition in their industry had attempted to form a co-operative selling organization – the Dominion Syndicate – to "take over and finance the pack of the companies, and to market it in such a way to stabilize prices" (Elder, 1986:3). Lack of co-operation, among other factors, led to the failure of the Syndicate, but it was not long before other forms of collaboration were to prove more durable. In 1903 the Canadian Canners Consolidated Companies Limited was formed to amalgamate seventeen companies. By 1910 this organization underwent a further metamorphosis, to form Dominion Canners Ltd., with the acquisition of a further seventeen independent canners. By this time Dominion Canners was reported to control over 80 per cent of the total Canadian output of canned fruit and vegetables. By 1915 an economic crisis brought on further moves to "rationalize" its operations, and Dominion acquired seven more canners in British Columbia. The pace of acquisitions quickened after 1923, when thirty-four independent plants joined the fifty plants Dominion already owned to form the new Canadian Canners Ltd. By this time the *Financial Post* was reporting, "Canadian Canners Ltd, has, by national expansion, purchase and merger, grown to such an extent that it occupies the dominating position among canners in the British Empire" (quoted in ibid., 30).

This kind of concentration of market power is not without its benefits for the firm that consolidates, or without its disadvantages for those who must sell to or buy from such a firm. Commenting on the large number of factories under Dominion's control, the Royal Commission report frankly stated, "The basic reason for this company's possession of so many separate units, many of which were closed as soon as bought, is its desire to gain control in the industry and remove competition" (Canada, 1937:68).

Indeed, even before 1906, long before the company had reached the zenith of its power, the *Annual Report of the Dept. of Agriculture of Ontario* noted the ability of the firm to pressure both its suppliers of manufactured goods (boxes, cans, sugar) and the wholesale houses it dealt with: "With its tremendous pack, it is able to control, or at least this is current opinion, 90 per cent of the wholesale houses in the Dominion under very strict instringents which render it a losing proposition should any wholesalers attempt to handle products of other concerns.... With these prevailing conditions it is obvious that the stronger the Canadian Canners get, the greater will their influence be in restricting competition and distribution" (quoted in Lockyer, 1983:89-90).

In the Canadian canning industry, a vital factor in securing a prominent position was a company's arrangements for the purchase of cans. Cans constituted, right after the raw agricultural product, the single largest cost of production. For instance, data from Canadian Canners' financial statements in 1930 showed that agricultural produce accounted for 35 per cent of total production costs, with cans the next major cost at 22 per cent. The third most significant factor, labour, was only 10 per cent of the total costs (Canada, 1937:70). Any firm that could significantly lower the cost of cans would have a competitive advantage over other firms. Canadian Canners was able to do this, in fact, through a special arrangement it made with the American Can Company, the largest of the two can manufacturers in the country. The arrangement provided Canadian Canners with a preferential discount on cans in return for a commitment to purchase cans from American Can for a period of twenty years. The discount meant that Canadian Canners had an advantage of 15 per cent in purchasing cans compared to its competitors (ibid., 71). In the words of the Royal Commission on Price Spreads, "It is reasonable to conclude that this advantage has a greater affect on the competitive position of Canadian Canners Ltd., than that which might be gained by any other exercise of the power that naturally accrues to the dominant manufacturer.... It provides a striking example of the monopolistic arrangements reacting to the injury of producer, wage earner and small competitor" (ibid., 71-72).

Not altogether surprising were the benefits this arrangement conferred on the American Can Company. Even with the onset of the Depression and the rapid downturn in revenues for many firms, the profits of American Can, after signing its "deal" with Canadian Canners, rose impressively from 9.9 per cent in 1931 to 21.1 per cent in 1933. Meanwhile, between 1929 and 1933 contract prices paid to growers by Canadian Canners dropped by some 40 to 50 per cent for crops such as tomatoes, peas, beans, peaches, and cherries (ibid., 154).

Nor were farmers immune to the extraordinary market power of Canadian Canners, judging by the accounts of farmers who attempted to boycott the firm by growing other crops to protest low prices, one-sided contracts, and the exorbitant dockage charges levied by the firm (ostensibly to compensate for produce that didn't meet its specifications) (Lockyer, 1983:90-91). Not surprisingly, farmers in some Ontario counties, such as Prince Edward, were known to have banded together to fight unfair practices and force a local factory to agree to take all of their crop, as company contracts left it entirely up to the factory manager to decide how much of a farmer's crop would be purchased.

The response of the agribusiness consortium to these farm protests was typically swift and, it would seem, successful. In some instances the companies counterattacked by importing produce from outside the region, including produce from the United States. Another company strategy used to counter farmers' initiatives was to begin negotiations to buy farms near each of the company's main canning factories in the county – although when this bluff failed to break the farmers' resolve the company representative was forced to go through with the actual purchase of farms to supply its plants, an action that did achieve the desired result of forcing farmers back into signing contracts they had protested against (ibid., 96-7).

About this time it was noted that to further reinforce its powerful position in the market, Canadian Canners maintained a practice of carrying over a substantial proportion of the previous year's production. This was apparently part of a policy of price-cutting for the purposes of restricting or eliminating competition. The Royal Commission on Price Spreads noted that this behaviour had a "disturbing and depressing effect" on the industry, making it "exceedingly unstable." It is perhaps not surprising that in the early 1930s growers producing eight principal crops for the firm received, on average, only 18 per cent of the final sale price of the products, and in one year the containers used for the goods made up a larger share of the products' final sale price (Canada, 1937:69,153).

Two Firms Dominate Meatpacking

The case of the Canadian canning industry was not simply a market anomaly: in the meatpacking sector a similar lack of competition also became notably evident during the Depression. The situation was of no small concern, due to the significance of meatpacking as the third-ranking manufacturing activity in the country in terms of gross value of output (it ranked first in terms of the materials used). It was an industry depended upon by a large portion of the farm community, given that livestock raising had come to be a central characteristic of much of the farm economy. Some 72 per cent of census farms in 1931 reported having non-dairy cattle, and 60 per cent reported raising swine.

Again, the comments of the Royal Commission on Price Spreads about this industry are telling. "The History of the past twenty-five years," the Commission's report noted, "is little more than a succession of mergers and consolidations, resulting in a steadily decreasing number of medium-sized plants, and an increasing dominance by the consolidated units" (Canada, 1937:55). In fact, two companies – Canada Packers and Swift Canadian – had come to play the preponderant role in the industry, together controlling 85 per cent of the $92 million Canadian meat industry by 1933 (ibid.). Of these two companies, Canada Packers was the larger, with sales almost double those of Swift. Significantly, in the United States the two largest packers, Swift and Armour, did not between them share as large a percentage of the industry's total production as Canada Packers alone accounted for north of the border (ibid., 59-60).

The significance of the monopolistic nature of the meatpacking industry was illustrated by the packing houses' ability to protect themselves from the ravages of the Depression, in comparison with other groups involved in the industry and primary producers in particular. Between 1929 and 1932, sales in the meatpacking industry had fallen dramatically, by some 50 per cent, or from $186 million to $91 million, due to a drastic shrinkage in demand. However, while returns to primary producers of beef fell by almost 57 per cent, the packing industry experienced a decline in revenues of only 24.5 per cent. The Royal Commission stated, "The packing industry has been able to better protect its margins than has the primary producer.... The manner in which these results have been achieved has a direct relation to the monopolistic character of the structure of the industry. The dominant position of the two largest companies, with extensive storage facilities and control of a great proportion of the slaughtering equipment in the country ... secured for them some measure of control over both livestock prices and selling prices for their product" (ibid., 56).

The prices these firms were paying livestock producers can in part be explained by the fact that by the 1930s the bulk of cattle shipments were going directly to the packing plants, rather than through public stockyards as they once had. Because the stockyards represented the only really competitive market, direct shipment to packers put the livestock farmer in a particularly vulnerable position. For one thing, all livestock delivered to the packers was weighed on the packers' scales, not on government scales with an impartial overseer. Secondly, in the case of beef at least, the packer alone determined the grade of the meat, and hence the price it would fetch. In a classic case of understatement the Royal Commission noted, "In these circumstances is it not difficult to see who is in the strong and who is in the weak position" (Canada, 1937:162).

When livestock still did go to the public stockyard, it was not at all clear that a state of real competition existed any more. A former official of Canada Packers gave "uncontradicted evidence" to the Royal Commission on this score, stating, "In Toronto it was the usual practice for this firm to arrange with Swifts, before the market opened, as to the prices to be paid for the purchase of livestock." Also the managers of the Western Stockgrowers Association, one of the largest ranchers in Alberta, gave evidence that if a packers' buyer gave an offer for cattle on the ranch, it would not be raised for any other packer-buyer either on the ranch or in the stockyards (ibid., 162).

It is perhaps most telling that, in a period of drastic shrinkage of the economy and widespread bankruptcy – personal and corporate – throughout the country, Canada Packers was still able to turn a profit even in the worst years of the Depression. "This relative prosperity of Canada Packers," the Royal Commission said, "bears some relation to the enjoyment of relative freedom from competition. That the inadequacy of such competition has operated to the detriment of the primary producer seems evident" (ibid., 57).

The development of a near monopoly in this key industry had other consequences. By diversifying into the cream and vegetable canning business, Canada Packers hoped to defray freight charges on railroad shipments of beef. Nonetheless, the presence of such a large firm in an industry characterized by relatively small competitive firms was destabilizing, particularly when Canada Packers got into the practice of offering its creamery products as "loss leaders" to chain stores.

A further consequence of the extraordinary control of the market for meat products was the ability of the largest packer to offer special prices to the emerging food-chain retail operations, to the detriment of the independent butcher shops. This practice is a clear example of how the largest processors provided extra benefits to the chain operators in the early stages of their growth, and thereby helped these chain operations further consolidate their own operations. In a later era, these chain-store operations would come to control a much greater portion of the retail food market and would then turn the tables and be in a position to dictate terms to even the largest of the food processors. It would thus appear that the logic of the concentration of capital is that once it consolidates in one sector of the economy it is led to foster concentration in related economic sectors by the very dynamic of accumulation.

The Tobacco Monopoly

Nowhere was the absence of any semblance of competition more apparent than in the Canadian tobacco industry. Nor was this an insignificant industry, with total production worth over two-thirds that of the meatpacking industry. The industry was massively dominated by the Imperial Tobacco Company, itself a subsidiary of a British firm, the British American Tobacco Company. Imperial Tobacco accounted for over 70 per cent of Canadian production in the early 1930s.

The onset of the Depression brought to the fore the ways in which unchallenged corporate mergers had prejudiced the position of the less powerful players in the industry. Farmer-suppliers to the industry were, as one might expect, the first victims of this situation. After examining the Imperial Tobacco Company's crop inspection books, the Royal Commission accused the

James Stanley McLean
1876–1954
First President of Canada Packers

James Stanley McLean was the first president of Canada Packers, a company he created out of four smaller ones. Little is known either about his personal life or about how he ran Canada Packers. He lived a secretive life out of the public eye.

McLean was born in Kendal, Ontario, but grew up in Port Hope. After completing high school he entered the University of Toronto as a math and physics student and later became a teacher. After two years he decided to leave teaching and change his career. He moved to Trail, British Columbia, and began selling insurance. He lasted a year at that before returning to Ontario. This time he ended up at the Harris Abattoir Company in Toronto as a junior clerk. McLean quickly moved up through the ranks. In 1905 he became secretary-treasurer, and he eventually became a partner.

During the First World War the beef industry in Canada was booming, with Canadian companies getting war contracts to supply beef to the allied troops in Britain. After the war the industry went into decline. The scale of the industry was too large for the domestic postwar market. Some beef-slaughtering companies had financial problems, and the banks refused them further loans. McLean saw the opportunity to merge some of the troubled companies with Harris Abattoir, which had not felt the same financial crunch.

company of taking deliberate advantage of a weakening of demand in the early 1930s to break the market. As a consequence, prices for tobacco had been forced sharply down to a point where growers were experiencing considerable distress. Profits in the processing sector remained consistently high, however, despite the conditions in the wider economy – although most of these profits were appropriated by the top one or two processors, leaving little for smaller manufacturers and retailers. As the Royal Commission noted, "The exorbitant profits that this company has been able to make, even in a

In 1927 McLean put together a new company that would last for sixty-three years and become one of Canada's largest food processors. He merged three meatpacking companies – Gunns, William Davies Company, and the Canadian Packing Co. – with Harris Abattoir. The merger created Canada Packers, and J.S. McLean was appointed the company's first president.

McLean remained as president for twenty-seven years and during his reign proved to be a controversial figure. He was outspoken and almost immune to any criticism. He was considered by some to be a "liberal industrialist" who believed that his workers were as important as he was to the company's success. He started a profit-sharing program with his employees and by 1942 was co-operating with the union. However, others regarded him as a "despot and the implacable foe of labour."

McLean was known as a shrewd businessman who was able to make a profit for Canada Packers even during the Depression. Another part of his success was to ensure that every part of the animal was processed and contributed to a profit. Other strategies of his caused criticism, such as filling empty spaces in cattle cars with his canned goods. This gave him a competitive advantage over the firms that did not run cattle cars.

McLean was undoubtedly one of the most successful and wealthiest men in the Canada of his time. He merged and created a company that held 59 per cent of the market share of the beef-packing industry in the 1930s. He also expanded the business to include edible oil products, creameries, and canned vegetables. In 1954, one month after his death, his son William took over the position of president. Almost four decades later the family's shareholdings would be sold to a British company that would completely change both Canada Packers and much of the food industry in Canada.

Source: Interview with McLean's son, William; Willis (1964); *Maclean's Magazine*, 1951:61.

period of general economic distress, are proof that a dominating position can be used to avoid the necessity of sharing in that distress" (Canada, 1937:52).

The largest processor used its market power not only to depress prices to producers, but also to maintain prices of tobacco products at artificially high levels and thereby disadvantage consumers. The Royal Commission described how Imperial Tobacco secured a system of price maintenance:

Their method of enforcement is simple. The Imperial Tobacco Company merely removes from its lists dealers and jobbers who cut prices, either of their own or competitors' products, with the jobbers' associations assisting by bringing the names of offenders to the notice of the Company. "Cutting off the list" in this case is no mere gesture. When a company which produces nearly three quarters of the supply refuses to sell to a wholesaler or retailer, the effect on that dealer is too obvious to need comment. (Canada, 1937:53)

The case established by the Royal Commission concerning Imperial Tobacco challenges any notion that the dominant position of this firm was achieved and maintained by dint of the superiority of its products in the marketplace. The Commission found that the company would not only employ such practices as threatening to cut wholesalers and retailers off its list, but also force dealers to sign price-maintenance agreements for its products. In addition, a host of complaints came to the Commission about other practices of Imperial Tobacco, including the claim that wholesalers were being compelled to push Imperial Tobacco products and hinder the sales of independent competitors, evidence that the advertising materials of smaller manufacturers were spoiled or destroyed, and contentions that Imperial made the supply of its more popular brands to retailers conditional on them taking less popular merchandise. The Commission concluded, "In its efforts to press the sale of its products, [Imperial Tobacco] has used its power in a way that is to be condemned and has indulged in unfair competitive practices" (Canada, 1937:54).

The Input Side

If through corporate concentration and restrictive practices the manufacturers that purchased the farmers' produce were protecting themselves from the devastation of the economic depression, often at the expense of producers, what can be said of the companies that produced the main inputs that had become essential for the carrying on of agricultural production? In the period before World War II, the use of agricultural chemicals had not taken on the importance it has today, and even the use of mobile mechanized power in the form of the tractor had not yet displaced the horse on the majority of Canadian farms. This was to take place largely after 1940. Nevertheless, the state of competition in the farm implements industry during the 1930s is pertinent, firstly because farmers had become dependent upon a wide variety of implements by that time, and secondly, because with the "tractorization" of Canadian agriculture that was to occur after 1940 farmers would become even more dependent upon a sector of agribusiness whose corporate structure was established in an earlier period.

Despite the variety of companies that had since 1860 or so emerged in Canada to supply farmers during the first technological "revolution" of Canadian agriculture, by the 1930s only four companies – Massey-Harris, International Harvester of Canada, Cockshutt Plow, and Deere and Company – controlled about 76 per cent of the sales of all agricultural implements in the country.[4] Moreover, because 17 per cent of the rest of the business was supplied by U.S. subsidiaries doing business in Canada, there was very little room left for smaller independent Canadian firms. Of the big farm-implement firms, the largest was the U.S.-based International Harvester, which controlled about 33 per cent of the market in the 1930s (Canada, 1971:39).

One of the consequences of a highly oligopolized farm implement industry can be seen in the relative inflexibility of the prices for their products, when compared to the stark declines in the incomes of primary producers during the Depression. After all, the average income of farmers dropped by 64 per cent for Canada as a whole between 1928-29 and 1932-33, and by 94 per cent in the Prairie provinces alone (Lipset, 1968:123, Table 10). Obviously, this decline would have occasioned a drastic restriction in demand, and a competitive supply industry would have had to respond with substantially lower prices. The Royal Commission found, however, that after four years of economic collapse, implement prices in 1934 were still some 90 to 96 per cent of those in 1930. It concluded, "The policy of the Big Three of the implement manufacturing industry since 1929 has been, in effect, to maintain prices and to adjust their production to sales at those prices, in other words, price inflexibility and production flexibility" (Canada, 1937:62).

CHAPTER SIX

CONSOLIDATION OF THE CANADIAN AGRO-FOOD COMPLEX

• • • • • • • •

They all threw in the towel. They didn't see the sense of losing a lot of money. They were forced by the wayside. They were the last of the independent vegetable processors.

Food industry observer on the plight of independent Canadian food processors, 1987

In a competitive system, survival depends on efficiency. In oligopolistic or monopolistic systems, survival depends on power – economic and political.

W. Heffernan and D. Constance, rural sociologists and observers of concentration in the food system, 1992

What we will have, if this march of increased concentration continues, is a national oligarchy in which a few dozen people will interact to bargain about the economic future of millions.

Robert Bertrand, former director of the federal Anti-Combines Branch, 1980

It is commonplace to think of World War II as one of the main turning points of the 20th century, and this is certainly true with the development of the modern food system. It was really only in the postwar epoch that the complex of activities related to the provision of food in modern industrial societies took shape. By this time the majority of the food produced in highly developed food economies was bought by privately or co-operatively owned corporations engaged in the transformation of food. In the United States, for instance, by 1975 some 66 per cent of the value of marketed farm production was bought by food-processing companies. Of the remainder, 19 per cent was exported in raw unprocessed form, and most of this was processed by food manufacturers outside the United States (Connor et al., 1985:23).

Discussion of the food-transformation industry brings to the fore several important social and political issues. One of these is the question of corporate mergers and the contention that the resulting high concentration of economic power causes inefficiency, waste, and inequality, both vis-à-vis primary producers and in the wider society. Another is the phenomenon of advertising – especially the vast resources of society it consumes and the question of its real value to society. Finally, the linkages between food processors and the producers of a variety of agricultural commodities illustrate the implications of the imbalance of power in the food system.

The Food System in Late-Capitalist Societies

In his work Louis Malassis identified a fourth stage in the development of the modern food economy, the one that characterizes our present system: the internationalized agro-industrial food economy (*l'économie alimentaire agro-industrielle internationalisée*).

Malassis' first and second stages correspond, for the most part, to the precapitalist era of world history. Modern capitalism, as a generalized mode of production, became firmly established in England by the 19th century and rapidly spread through Western Europe and North America at that time. Nevertheless, significant residues of the second stage, the "agricultural and domestic food economy," persisted in Western Europe, and certainly in Eastern Europe, until well into the 19th century. Malassis' third and fourth stages are specific to the capitalist era. They are associated with the continuous technological change that together with the separation of people from the land and from other means of producing a livelihood lay at the core of the industrial revolution.

It is only in the fourth state that we see the maturation of what is a thoroughly capitalist food system in all respects *except* that of agricultural production. The sphere of production remains predominantly in the hands of increasingly capitalized, but not "capitalist," family units of production.[1] With the maturing of the capitalist food system we see the full development of the various branches of the system that had emerged in the previous stage.

The Seven Subcomplexes of the Food Economy

Following Malassis and Padilla (1986:185), and adapting their schema to the Canadian situation, we can designate the agro-food complex as the ensemble of activities that organize the provision of food in a given society. In reality, several subcomplexes of activities exist within a major food economy.

i) *Agriculture and fishing*, that is, the production or securing of the basic foodstuffs destined for transformation and distribution regionally, nationally, and increasingly internationally.

ii) The *food-processing industries*, producing both intermediate goods (flours, sweeteners) and final consumption goods. This subcomplex can be further broken down into a capitalist sector, made up of both private companies and public joint-stock holding companies on one hand and the farmer and fisher-controlled co-operative processing sector on the other. In several Canadian provinces, co-operative enterprise has a dominant market position for key activities such as the dairy industry, fruit processing, and even livestock and poultry processing.

iii) *The wholesale distribution and retail sector.* In Canada this segment of the food economy is, for the most part, now tightly integrated under the control of the giant retail food chains.

iv) The *institutional food industry*, which provisions an increasingly wide array of institutional environments, including schools, universities, corporate cafeterias, hospitals, and government offices. Certain agro-industrial firms may be specialized solely to supply this sector.

v) The *input and service industries*, which provide other segments of the agro-food complex with a wide range of intermediate goods or inputs. These include, in the case of agriculture, farm implement manufacturers, firms producing fertilizers and ag-chemicals, seed companies, feed companies, and veterinary supply companies. In the case of the processing industry, this includes inputs in the form of containers (cans, bottles, boxes, plastic bags), chemical food additives, and processing equipment. Thus the agricultural and transformation (processing) segments of the agro-food complex provide essential demand to a very large and complex group of independent enterprises, or subsidiaries of larger corporations that depend upon the agricultural and processing segments for a market.[2]

vi) The various *state and quasi-state institutions, apparatuses, and regulatory structures* that are characteristic of the food economy. The specific features of these state and quasi-state institutional arrangements vary considerably from society to society. At the most general level, they are an outcome of the protracted struggles of competing class interests within society, struggles that typically span a considerable time period. In the Canadian case, these institutions range from state agencies such as the Canadian Wheat Board, which has for decades organized the sale of the most significant of Canadian export crops (a task dominated by the private sector in other societies), to apparatuses having more independence from the state – such as producer marketing boards governing the production and sale of fruits and vegetables that go for processing. The influence of marketing boards – and this varies from province to province – is determined nevertheless by the powers conferred on them by provincial government legislation.

A further institution that is vital for the reproduction of our food system is that of higher education and research – principally certain universities and agricultural colleges – which have taken on responsibility for the formation of each new generation of agricultural producers, food engineers, technicians, scientists, and management. These institutions also organize the development and dissemination of new technical approaches and innovations throughout the food system.

For the most part these institutional structures that have defined how our food economy has evolved were the outcome of strong pressures from agriculturalists in the early part of the century. In many cases these producers were attempting to protect themselves from the rapid increase in market power of major capitalist firms in the banking, transportation, grain storage, and processing spheres of the economy. The countervailing power of the capitalist sector has in turn tirelessly worked at resisting the demands of farmers and their political organizations. This interplay of pressures emanating from conflicting class interests within, and outside, the food economy is now once again coming to the forefront as a powerful array of forces attempt to dismantle at least part of this institutional structure (a matter considered in greater detail in chapter 8).

Several distinct processes distinguish the fourth stage of the food economy from the third: new forms of corporate concentration with the development of conglomerate and diversified food corporations; the multinationalization of capital employed in this economic sphere, and the influence of foreign ownership within our food system; the central role of mass-market advertising in establishing *product differentiation*, and the phenomenon of product proliferation and its consequences; and the growing integration of farming and agribusiness operations, and the emergence of food retail chains as the leading players in the food system.

Corporate Concentration in Comparative Perspective

Karl Marx saw the concentration and centralization of capital as one of the core tendencies of capitalist economies (1977:582-89). The history of the food system in the 20th century has certainly borne out his prediction. In recent decades the further concentration of economic power in the food system has created new forms of corporate organization that, in turn, pose new problems for society at large.

In the second half of the 20th century the food sectors of advanced capitalist countries have been the site of substantial merger activity, with high rates of corporate concentration the result. In the United States, for example, by 1962 the top fifty processing companies controlled 70 per cent of sales in almost all product categories (Frundt, 1981:27). In the United Kingdom, which for a variety of reasons also has a highly concentrated food business, thirty companies accounted for three-fifths of employment and value added in the food industry by the late 1970s (Burns, 1983:3).[3] Indeed, the top *three* firms had about 70 per cent or more of sales in each of the following product categories: breakfast cereals, bread, canned soup, flour, ice cream, potato chips, sugar, and tea (Howe, 1983:105).

The food and beverage processing industry in Canada, which has been the largest sector of manufacturing in the country in terms of shipments and employment (Canada, 1981:46), has not been immune to the changes that have restructured the industry elsewhere. By 1968, for instance, a federal government report had noted that the food industry was one of the most concentrated sectors of Canadian industry (OECD, 1979:185). A report examining data in the 1970-80 period shows that at the national level the concentration (as defined by the ratio of shipments by the four largest enterprises in the industry to shipments by the industry as a whole) continued apace in most subsectors of the Canadian food-processing industry. Companies in this sector have reached a markedly higher degree of concentration than have similar companies in the United States (Veeman and Veeman, 1978:764). The only subsectors experiencing a slight decline in concentration were the slaughtering/meat-processing and distillery subsectors (Loubier, 1984:8,27).

The high degree of concentration of the food industry in Canada is not unique to this sector. Indeed, several other sectors, including petroleum, petrochemical, airlines, and telecommunications, are even more tightly dominated by a few firms. Using a comparative perspective, we can see that the levels of corporate concentration in Canada are truly remarkable, noticeably higher than leading industrial powers such as the United States and Japan (see

Figure 2
Share of Total Non-Financial Activity Held by the 100 Largest
Non-Financial Corporations

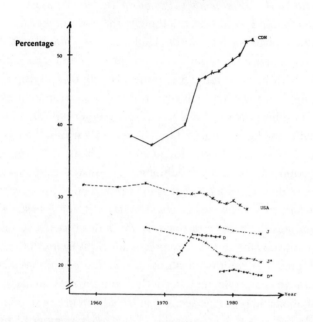

Legend: CDN: Canada (assets); USA: United States (assets); J: Japan (assets, incl. domestic subsidiaries); D: F.R. of Germany (sales, all firms); J*: Japan (assets, excl. subsidiaries) D*: F.R. of Germany (value added, all firms)

Source: Marfels (1988:82).

Figure 2). This level of concentration is not altogether surprising given the inadequacy of Canada's legislation aimed at ensuring competition in the marketplace. As the Report of the Royal Commission on Corporate Concentration put it, "Enforcement of the Combines Investigation Act has been ineffective in discouraging mergers that increase concentration within an industry" (1978:111).

The structure of the food economies in all the advanced capitalist countries exhibits a considerable degree of variation, however. For example, by the 1970s France and Italy still had small craft-like food-processing establishments with fewer than five employees forming 33 per cent and 29 per cent of all food-manufacturing establishments respectively, as compared to only 1.3 per cent for similar food establishments in the United States (OECD, 1979:135).

For much of the postwar period food corporations in the United States, and to a lesser extent in the United Kingdom, have set the trends worldwide. The expansion of these firms has entailed several processes, including the development of multinational operations, diversification, and the emergence of conglomerate firms. The *multinationalization* of the food industry started largely after the Second World War, although the forerunners of today's processing multinationals, such as Nestlé, Unilever, and United Fruit, began establishing foreign subsidiaries early in the century. The major period of multinational growth began after 1955, spurred largely by U.S. capital and later by U.K. firms redeploying their investments into continental Europe (OECD, 1979:274). This growth proceeded in the manner typical of postwar expansion in home markets – largely through takeovers within the food sector rather than through the establishment of new plants. This process of multinationalization has been overwhelmingly dominated by U.S. capital, with 68 per cent of the largest one hundred food-processing firms in capitalist countries with foreign subsidiaries being American.[4] Capital based in the United Kingdom was the next largest participant in this process, accounting for 32 per cent of the one hundred largest processing firms having foreign subsidiaries. In this process of multinationalization, with few exceptions (Canada Packers, McCain, Seagram, and, more recently, John Labatt) Canada has had few firms in the food and beverage sector that have been in a position to participate. Canada has largely been the site of the redeployment of foreign (U.S.) capital.

The concentration and centralization of capital and the multinationalization of enterprises are, of course, not unique to the food industry, but the food industry provides a rich ground from which to consider the profound implications of this process. Indeed, a critical study of the wider consequences of contemporary trends that have reshaped the corporate structure of the Canadian food economy raises pertinent and pressing questions about the current organization of the wider Canadian economy.

The Conglomerate Firm, Cross-subsidization, and the Demise of Competition

While it is important to look at the consequences of having one or just a few corporations dominate a specific industrial sector in the food system, the reality today is increasingly more complex. As the English economist Joan Robinson noted some time ago, "More and more, the great firms have a foot not only in many markets, but in many industries, in several continents" (quoted in Connor et al., 1985:242). What this refers to is the rise of the *conglomerate* firm, a relatively little-known phenomenon of industrial organization. There is evidence, much of it from the United States, to suggest that by means of conglomerate structures corporations can markedly enhance their market power and seriously outcompete more specialized firms. Because conglomerate firms characterize the Canadian food economy, we need to seriously consider the implications of this phenomenon.

Jorge Niosi (1981:47) describes conglomerates as "companies or groups of companies under unified control that produce goods and services in unrelated sectors of activity." One of a conglomerate's most important features giving it such potential market power in a given industry is the possibility it has for engaging in what economists call *cross-subsidization*. This is the practice of using profits gained in one activity to subsidize their entrance into other markets (see Connor et al., 1985:243). In other words, when a large conglomerate firm enters an industry where its sales are small relative to its entire operations, it can outspend its smaller rivals and sustain losses for a period of time, which is usually enough to drive its smaller competitors to the wall. Once the competition is out of the way, the firm is free to charge higher prices, which will eventually make up for the loss incurred in securing the dominant market position.

This is not merely speculation about the *possible* behaviour of conglomerate firms; there is evidence to indicate that this is precisely what conglomerate firms have done. Much of that evidence comes from the United States, in large part because of the more stringent laws regarding corporate disclosure there, along with the fact that compared to Canada more vigorous anti-trust legislation periodically makes public what are normally very sensitive details on corporate behaviour.

One of the more dramatic examples of the significance of the cross-subsidization of conglomerate firms in dramatically altering the competitive environment of the food economy was the entrance of the giant grocery conglomerate Procter & Gamble into the retail coffee market. Even by the early 1980s P&G's aggregate sales were approximately $21 billion, making it one of the largest food-related companies in the world.

Procter & Gamble's strategy for capturing a large market share in the U.S. coffee market began with its purchase of an already existing coffee company, J.A. Folger, which had an important market share in the western United States. In the early 1970s, P&G launched its campaign to control a significant part of the eastern U.S. market. At that time this market was dominated by another large conglomerate food company, the General Foods Corporation, owner of the Maxwell House brand. As P&G launched its promotional campaign in city after city in the East, a massive retaliation was mounted by General Foods to protect its market share. Utilizing a combination of generous mail-out coupons to consumers, in-pack coupons, trade allowances, and display allowances to retailers, General Foods counterattacked P&G's campaign in one Eastern city after another. It then drew on its tremendous revenue base to bring its countercampaign to P&G's home markets in the Western states.

While the attempt by P&G to grab a substantial part of the U.S. coffee market from General Foods could be viewed as merely a colourful example of two corporate giants fighting it out head to head in the marketplace, this would be to miss the point. In fact, this corporate battle had serious consequences for a host of other players as well. While the war in the retail coffee market was basically a sideshow for the two conglomerate contenders, who could count on revenues from other sources to weather the storm, it proved to be devastating for the remaining regional and local coffee-roasting firms, who tended to be specialized in this commodity alone. Unlike General Foods, these smaller firms did not have a war chest of retained earnings they could use to defend their markets from the juggernaut advertising campaigns of the conglomerate giants.

As John Connor, Richard T. Rogers, Bruce W. Marion, and Willard E. Mueller note in their landmark study of the U.S. food industry, one leading local firm in the Pittsburgh area saw its market share drop from 18 per cent to 1 per cent as a result of this coffee war between the two grocery giants (1985:263). A Syracuse-based family-run firm suffered an 80 per cent loss in its retail sales when P&G swept into its market area. These stories were repeated in city after city, and in the end the number of coffee-roasting companies in the country dropped sharply and, we can assume, so too did any effective competition in the marketplace. As Connor and his co-authors conclude of this episode: "P&G-Folger's behaviour and General Foods' response to it illustrate how conglomerate firms may destroy smaller, *though not necessarily less efficient*, competitors. Such strategies are destructive of competition when a huge conglomerate takes aim at one product or market at a time, destroying less powerful competitors by cutting prices and/or elevating costs through massive advertising and promotional campaigns. These practices usually involve subsidising losses in one market or product from profits earned elsewhere" (1985:265).

This episode had a fairly immediate impact on consumers. As world coffee prices dropped dramatically in the late 1970s, consumers wondered why this decrease was not being reflected at the retail level. Behind the scenes, the war in the retail coffee market between the two conglomerate giants had resulted in an expansion of the market share for each firm, largely at the expense of the smaller regional coffee roasters. Having removed much of the competition, the two conglomerate corporations experienced less pressure to lower retail prices once the wholesale price fell. Rather, as *The Wall Street Journal* suggested on October 7, 1981, the big players used their enhanced earnings to expand their promotional war chests.

The devastation caused by the predatory behaviour of conglomerate firms in the retail coffee market was not an isolated incident. Detailed evidence shows how the U.S. brewing and baking industries were dramatically restructured with the entrance of the conglomerate firms Philip Morris and the International Telephone and Telegraph Corporation into those respective industries. Through massive advertising campaigns and predatory pricing policies, the firms were eventually able to establish their dominance within a new industry, despite long-established rival firms that were often more efficient producers.[5] Connor, Rogers, Marion, and Mueller conclude that when it comes to conglomerate firms, "Success or failure in this environment is not determined by efficiency but by sheer economic power" (1985:265).

Conglomerates and the Canadian Food System

Historically, conglomerate companies such as Canadian Pacific have left an important mark on the structure of the Canadian economy. Even as early as the late 19th century, CPR was not only the largest private enterprise in Canada, but also the first to achieve a conglomerate structure, with interests in railways, shipping lines, hotels, and real estate. Today, large conglomerate firms are very much a part of the Canadian economy. This is not surprising given that the legislation in Canada dealing with the corporate mergers or acquisitions that tend to restrict competition – the Combines Investigation Act – has not had the power to deal with combinations of firms in several industries (Canada, 1978:103).

In the 1970s the Royal Commission on Corporate Concentration in Canada noted that there were already more than thirty conglomerate firms in Canada among the top two hundred publicly held non-financial firms (Goldenberg, 1984:49). Top corporate managers saw the move to diversify holdings and develop a conglomerate structure as a way to bring greater stability to a corporation with activities previously confined more or less to one sector, such as brewing or transportation. A more diverse mix of corporate holdings would help even out cyclical ups and downs that affected the firm's primary sector of operations, or boost its profits if its primary arena of investment was plagued by slow or stagnant growth. Among Canada's largest conglomerate firms are a few giants controlled through a holding company, which in turn is controlled by one family or just a few individuals. Brascan Ltd. (Edward and Peter Bronfman), Argus Corporation (the Blacks), and Power Corporation in Quebec (Paul Desmarais) are some leading examples (see Niosi, 1981:ch.1,2).

Given the prominence of conglomerate firms in defining the corporate structure of the Canadian economy, it is not surprising that conglomerates control substantial segments of our food system. Two of the most prominent food companies, George Weston and John Labatt Foods, are part of extensive corporate empires. John Labatt Foods is part of the Edper Bronfman group, the Canadian conglomerate with the largest number of companies under its control: 421 as of 1992.[6] Weston and Edper are, in turn, ultimately controlled by single families that are at the pinnacle of our corporate elite. Figure 3 illustrates the extensive holdings of one of these prominent conglomerate firms, holdings that go beyond the food-related companies it owns or control.

The growth pattern of John Labatt Foods followed a pattern similar to that of the conglomerate firms entering new territories in the United States. Fuelled by revenues from the lucrative brewing industry, Brascan's Labatt division was able to start from a position of having very little exposure in the food sector and then, through rapid and extensive acquisitions, position itself on the road to becoming "a major North American food company" (*The Globe and Mail*, Sept. 11, 1992). Its milk-processing subsidiary, Ault Foods, for instance, bought up so many dairies that by 1983 it had nearly 50 per cent of the market and was attracting nervous attention from the federal government's competition bureau (ibid.).[7] In Canada only an extraordinary degree of control in the marketplace attracts this kind of government attention. Labatt's foray into the food business is also a good example of how bigness in the marketplace is so often associated with a lack of financial acumen or even basic market "savvy." While clearly Labatt had a lot of spare cash to throw around during its takeover binge in the food industry, both in Canada and the United States, a number of these businesses "drained millions and were eventually sold" (ibid.). By the early 1990s Brascan's Labatt subsidiary was busy trying to sell off most of its extensive holdings in the pasta, juice, frozen foods, and dairy businesses (*The Globe and Mail*, May 14, 1991, Sept. 11, 1992, March 12, 1993). While this move was being presented to shareholders and the public as a strategy to "get back to our roots" (that is, back to the brewing and entertainment industries), the reality was that this conglomerate had made a foray into the food business that had required an investment of hundreds of mil-

Figure 3
The Edper Bronfman Group: A Major Canadian Conglomerate

(Edward and Peter Bronfman)
Hees International Bancorp Inc.
Brascan Holdings Inc.
Brascan Ltd.
(assets in 1990 • $5.7 billion)

Natural resources
Noranda Inc.
(profits in 1989, $252m)[2]
• Brunswick Mining & Smelting (65%)[3]
• Brenda Mines (69%)
• Falconbridge (50%)
• Kerr Addison Mines (51%)
• Hemlo Gold Mines (51%)
• Noranda Aluminum Inc.
• Norandex Inc. (90%)
• Wire Rope Industries (90%)

Noranda Forest Inc.
• Fraser Inc.
• Island Paper Mills
• James Maclaren Industries Ltd.
• Normick Perron Inc.
• MacMillan Bloedel (50%)
• Northwood Pulp and
  Timber (50%)

Noranda Energy Ltd.
• Canadian Hunter
  Exploration
• Norcen Energy
  Resources (34%)
• North Canadian
  Oils (50%)

Consumer & industrial products
John Labatt Ltd.[1]
(profits in 1989, $135m)
• Labatt Brewing Co.
• La Brasserie Labatt
• Oland Breweries
• Labatt Importers
• Labatt Brewing U.K.
• Latrobe Brewing Co.
• Birra Moretti (77.5%)
• Prinz Brau ((77.5%)

John Labatt Foods Inc.[4]
(profits in 1989, $106m)
• Chef Francisco
• Omstead Foods
• Delicious Foods
• Oregon Farms
• Pasquale Food Co.
• Everfresh Inc. (sold in 1992)
• Ogilvie Aquitane (50%)
• Ogilvie Mills (sold 1992)
• Ault Foods
  • Ault Dairies
  • Royal Oak
  • Lactantia Ltée.
  • Ault Foods U.K.
• Johanna Farms
  • Tuscan Industries
  • Lehigh Valley Dairies
• Woodstone Foods
• Halls European Foods
• Canadian Pizza Crust
• McGavin Foods (60%)
• Catelli-Primo (46.2%)
• Canada Malting (19%)

Financial services
Trilon Holdings Inc.
(profits in 1989, $218m)
• Trilon Financial Corp. (40.3%)
  • Royal Trustco (48%)
  • London Insurance Group (58%)
  • Royal Lepage (52%)
  • Triathlon Leasing (75%)

• Labatt Entertainment Corp.
• BCL Entertainment (45%)
• Supercorp Entertainment (50%)
• The Sports Network
• Toronto Blue Jays (45%)
• International Talent Group (50%)
• Dome Productions
• Le Réseau des Sports (70%)

Foreign operations
Brascan Brazil
(profits in 1989, $43.0m)
• Brascan Imobiliaria
• CESBRA
• Banco Capitaltec (53%)
• Capitaltec Consultoria (53%)
• Capitaltec Futuros (53%)
• Ticket Servicos (40%)

Energy related operations
Great Lakes Group Ltd.
(profits in 1989, $103m)
• Great Lakes Power
• Unicorp Canada (24%)

Notes
1. Edper sold control of John Labatt Ltd. to investment dealers in February 1993 (*The Globe and Mail*, Feb. 13, 1993).
2. Profits for each division in millions of dollars.
3. Figure in brackets indicates percentage of stock owned, where known.
4. Part of the food division was sold in 1991.

Source: *Financial Post* Yellow Cards, 1990, 1991.

lions of dollars, and in the process it had apparently lost a very large sum of money. As a "reward" for getting their firm back on track, the conglomerate's two top executives were planning a special dividend for shareholders that, incidentally, would net each of them approximately $500,000 in 1992. It is unlikely that anything other than large conglomerate firms with extraordinary market power could afford to engage in such behaviour and still find themselves in business at the end of it.

In more general terms, it is unfortunate that little research is available to document the impact of conglomerate firm behaviour in Canada, as has been done for the United States. However, in objective terms we can say that there is, at the least, a very great potential in Canada for the predatory behaviour and anti-competitive outcomes that we know have characterized conglomerate firm behaviour south of the border. As one authority of the Canadian corporate scene has noted, "The markets in which these conglomerates operate are not competitive. The striking characteristic is shared monopoly, and with it, control of markets, regulation of production, control of the supply of raw materials and energy sources, and fixing of prices" (Veltmeyer, 1987:41). At this point, firmer conclusions on the impact of conglomerate firms on our food system await more empirical study of the problem.

Concentration and What We Pay for Food

The concentration of food manufacturing firms is linked to ever greater expenditures on advertising and other selling activities, but the level of corporate concentration in our food system has other serious consequences, as well. Perhaps the greatest knowledge of this comes with the breakfast cereal industry, largely as a result of anti-trust proceedings mounted against the major firms in the United States during the late 1970s. Since the subsidiaries of these U.S. manufacturers also dominate the Canadian market, it is a relevant case study for our own situation.

The breakfast cereal industry has been described as a tight oligopoly and one of the most highly concentrated industries in the U.S. manufacturing sector (Scherer, 1982:195). Even by 1970 the leading four firms, Kellogg, General Mills, General Foods, and Quaker Oats, controlled 91.5 per cent of sales. And while it is often argued that a certain degree of concentration is necessary to maximize a firm's economies of scale, the largest breakfast cereal firms are far larger than they need to be to realize maximum efficiencies from an engineering standpoint. Kellogg is an estimated seven to ten times as large as it needs to be to realize the principal production economies of scale (see ibid., 197).

These firms have not pursued a strategy of expansion and concentration to achieve economies of scale so much as to reap the advantages that bigness provides in the marketplace. A key advantage that oligopoly confers is in the area of pricing. Evidence suggests that the breakfast cereal industry has moved a long way from a situation of price competition. Rather, a regime of price discipline appears to have taken over, with the leading firm, Kellogg, taking the role of price leader in the industry. What this means is that Kellogg set the pattern in changing prices – almost always upwards – and the other firms followed. Frederic Scherer's study of the industry noted: "Out of 15 unambiguous price increase rounds between 1965 and 1970, for which period the documentary evidence is reasonably complete, Kellogg led 12. Kellogg's price increase was followed nine times by General Mills and ten times by General Foods; on only one occasion did neither follow.... *Leadership was sufficiently robust to permit price increases in times of both booming and stagnant demand*" (1982:203-4, my emphasis).

Through the mechanism of a price leader, then, these firms were able to avoid much of the price competition that is supposedly at the core of a "free market" system. Of the 1,122 price changes in the industry between 1950 and 1972, only 1.5 per cent were list price *reductions*, and half of these occurred in a single incident (ibid.). The demand for breakfast cereals showed little growth through much of the period. This price behaviour has little to do with what would occur if a free market were in effect.

Maintaining the breakfast cereal industry as an exclusive club has been rewarding for the few firms involved. The after-tax returns for the top five U.S. breakfast cereal manufacturers was a comfortable 19.8 per cent for the period 1958-70, at a time when the comparable figure for all manufacturing was only 8.9 per cent. And while this may make the industry an attractive investment opportunity for those with the money to invest, the extraordinary profits represent a social cost for society at large. To put it in the somewhat obtuse and cautious language of the economist, "Persistently high profits that are not attributable to unique or nonreplicable efficiency or resource endowments reflect resource misallocation and the unnecessary redistribution of income from consumers to investors" (ibid., 211). In other words, the absence of competition means that people are paying more than they would be if the market was in fact working.

In fact, in the 1970s the U.S. Federal Trade Commission stated that the overcharge by the few firms in this one sector of the food industry was costing Americans $100 million a year (*Consumer Report*, February 1981:76). One can only wonder what the cost of uncompetitive pricing for consumers is if we look at the food industry as a whole, rather than just one product classification. Such estimates are difficult to make without the benefit of documentation that is only rarely available, as in the case of anti-trust suits, or occasionally through Royal Commission inquiries. While I am not aware of an attempt to quantify this type of situation in the Canadian context, in 1981 the influential U.S. publication *Consumer Report* cited a study estimating that consumers in the United States were being overcharged about $15 billion as a result of excessive concentration in the food-manufacturing industries (ibid., 76).

Concentration, Advertising, and Social Waste

After the Second World War there was a rapid expansion of name-brand processed staples and a shift to more highly processed foods in general (cake mixes, breakfast cereals) with a higher value added component. As Figure 4 illustrates, the spread of processed foods has proceeded at different rates in different countries, and in some developed countries, such as France and Italy, social and cultural barriers have helped to slow the pace of this phenomenon. The development of branded staple products was predicated on massive advertising expenditures to maintain and strengthen product differentiation. Those firms able to establish successful branded products took a favoured position in the new postwar food economy, both relative to other food manufacturers and relative to food retailers that were becoming more and more concentrated and powerful (see Connor et al., 1985:79).

In the food business, then, ever greater expenditures on mass advertising have become absolutely central to the survival of brand products in the marketplace. While this phenomenon really took off in the 1950s and 1960s, mass advertising had been pioneered by capitalists in the food business in the early part of the century. W.K. Kellogg, the U.S. breakfast cereal entrepreneur, for example, was a major innovator in the use of mass advertising techniques in the first decades of the century. To create demand Kellogg committed a third of the working capital of his new company to a full-page advertisement in the *Ladies' Home Journal*, offering a season's supply of his corn flakes to any woman who persuaded her grocer to stock the Kellogg's product (Scherer, 1982:192). Between 1906 and 1940 Kellogg spent $100 million – an extraordinary sum for the time – to promote his product. It was largely through advertising that Kellogg was able to eventually push aside the host of rivals in the fledgling breakfast cereal trade (Powell, 1956:140-42).

Once television became the most effective marketing method, the largest

Figure 4
Consumption of Processed Foods, by Country

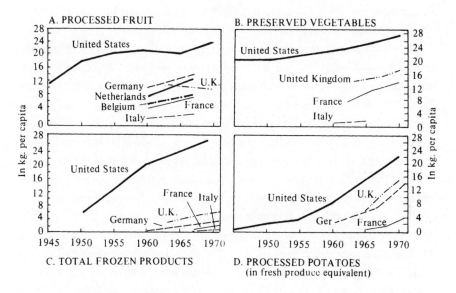

A. PROCESSED FRUIT

B. PRESERVED VEGETABLES

C. TOTAL FROZEN PRODUCTS

D. PROCESSED POTATOES
(in fresh produce equivalent)

Source: OECD (1979:118).

food firms found themselves in an especially advantageous position, notably because of the high cost of gaining access to mass media outlets. By the early 1990s a one-minute time slot during prime-time on U.S. national network television would cost $100,000 (*The Globe and Mail*, July 13, 1992:A12). Large firms were able to achieve substantial economies in buying advertising time and space, compared to smaller firms, and were thus able to further enhance their market positions. In fact, in the United States the ascendance of just a few firms to a commanding position in the food economy is confirmed by data showing that of the 1,100 food companies using major media outlets to advertise their products in 1982, only 12 firms accounted for 45 per cent of the advertising expenditures. This privileged access to the mass media enjoyed by a few large food corporations has probably been the most important factor in protecting their competitive position in the product markets they dominate (see Connor et al., 1985:ch.3).

In the United States the expenditures of the food firms on advertising as a proportion of sales are greater than those of corporations in virtually any other sector of the economy. Expenditures in this sector had for almost thirty years exceeded those of any other broad manufacturing category, and they have been growing in real terms as well (Connor et al., 1985:80-82). The same would appear to be true of Canada: of the top fifteen corporations advertising in the mass media, eight were engaged in the food and beverage industry (Nielsen Media Services, 1991).

Direct advertising in the mass media only accounts for a portion of the total cost of the food industry's sales effort in capitalist societies. There are substantial other costs that go into the successful marketing of food, especially in the case of the large companies. For example, it has been estimated that the so-called "pull" advertising devices – premiums, package design, trading stamps, contests, sweepstakes, and free samples – increase sales promotion costs in the United States by at least 65 per cent *above* media advertising costs (Gallo, 1981). There is little reason to think that Canada is much different in this regard, given the prevalence of the same marketing techniques here.

To these promotional costs must be added a whole range of expenses related to the sales effort directed by manufacturers towards food retailers to persuade them to stock their brands. These include fairly visible expenses such as direct field-sales forces, "point-of-purchase" displays for stores, and trade fairs and conventions. It also entails very significant but more or less invisible promotional practices such as "special deals" and allowances that processors regularly pay retailers to secure shelf space in their stores. It is notoriously hard to obtain hard data on the magnitude of this expenditure, because of the illicit nature of much of it. Nevertheless, in 1987 it was estimated that these hidden expenditures amounted to over $2 billion of the $32 billion of annual food sales in Canada (Matas, 1987:1).

The total economic resources that are included in the selling of food and beverage products are large indeed. Again, we have as yet only estimates of this from the United States, where John Connor and his colleagues write that total selling costs for the promotion of food and beverages alone were in the range of $10 billion to $15 billion in 1980. Although such information is less readily available in the Canadian context, we do know that in this country, of the top fifteen corporate advertisers in major media outlets in 1991, eight of them were engaged in the food and beverage sector. In that year, these eight firms together spent almost $350 million in Canada promoting their products on television and in magazines and daily newspapers (interview with Nielsen Marketing Research, 1992). Of the top one hundred advertisers in Canada, thirty were corporations engaged in the food industry, and these firms spent a total of $671 million on advertising in major media outlets in 1991 (see Table 2). If we can take as valid for Canada the estimate of Connor, Rogers, Marion,

Table 2
Ad Expenditures of Top 30 Food and Beverage Company Advertisers, 1991

	Expenditures (in millions)	% of all expenditures
1. Daily papers	$104,656.1	16%
2. Magazines	44,627.2	7%
3. Out of home	14,564.2	2%
4. Total TV	507,210.4	76%
Total	$671,057.9	100%

Source: Nielsen Media Services, National Ad Expenditure Annual Summary, 1991.

and Mueller that only 30 to 40 per cent of the total selling expenses of U.S.-based food companies are accounted for by mass media advertising, then it would appear that the top thirty advertisers in the Canadian food and beverage industry are spending, by themselves, in the order of $1.4 to $1.9 billion to promote their products on an annual basis.

One obvious question is: *why* has such a tremendous selling apparatus become integral to the food system? Clearly it was not always so. In an age when the food industry was on a smaller scale, with a much larger number of competing firms of similar size, advertising and the rest of the sales effort played a much less important role. This is what economic theory would lead us to expect: "Under conditions of atomistic competition, when an industry comprises a multitude of sellers each supplying only a small fraction of a homogeneous output, there is little room for advertising by the individual firm. It can sell at the going market price whatever it produces; if it expands its output, a small reduction of price will enable it to sell the increment, and even a small increase of price would put it out of business by inducing buyers to turn to its competitors who continue to offer the identical product at an unchanged price" (Baran and Sweezy, 1966:116). However, once the process of corporate concentration has dramatically reduced the number of sellers in an industry, as in our modern food system, competition around price tends to recede into the background, to be replaced by competition among firms through variation of the product's appearance and packaging, "planned obsolescence," and especially mass advertising to reinforce product differentiation. For as Paul Baran and Paul Sweezy have noted, the more successfully a company can differentiate its products from those of competing firms, the more this company is in the position of a monopolist in the marketplace, and the more able to extract extra profits: "The stronger the attachment of the public to [a firm's] particular brand, the less elastic becomes the demand with which [the firm] has to reckon and the more able [it] is to raise [its] price without suffering a commensurate loss of income" (1966:116).

In addition to brand differentiation, a related strategy pursued by oligopolistic firms to protect their market position is *product proliferation*. This too helps explain excessive expenditures on advertising in the food industry. Where at one time leading firms in the market resorted to predatory pricing wars to discourage new competition in an industry, in recent times firms, and especially those in the food industry, have favoured the continuous introduction of new grocery products as a more effective strategy to maintain their positions, rather than upsetting the price structure in a product category (Roland H. Koller, cited in Connor, 1981:611). This product proliferation creates powerful *barriers to entry* in the marketplace. This is because to survive a new firm has to capture a certain minimum market share in a given product category, and the more fragmented the market becomes the harder it is for new entrants to achieve the necessary share of the market without extraordinary expenditures on advertising. In the breakfast cereal industry, for example, it has been estimated that a new firm needs between 3.5 and 5 per cent of the market to make a profit. The U.S. Federal Trade Commission found that a new firm would have to introduce six to ten new products to earn a total market share of 3 to 5 per cent. The cost of doing so is very high. In the late 1970s advertising costs alone were an estimated $20 million to $35 million to achieve this objective (*Consumer Report*, February 1981:80). This high cost of entry has acted as a powerful deterrent for new competition in the industry.[8] It should be no surprise that the cereal industry continues to be dominated by the same few firms.

In an oligopolized industry, leading firms pursue product proliferation to protect their positions from potential competition. As a 1967 strategy statement prepared by an advertising agency for the leading breakfast cereal manufacturer, Kellogg, noted, "Continual proliferation of the pre-sweet [children's cereal] business through consecutive new product introductions is the essence of competitive strategy" (*Consumer Report*, February 1981:77). Not only does this strategy keep a market from becoming competitive, but it has also been said to be a major cause of excess advertising (John A. Henning and H. Michael Mann, cited in Connor, 1981:616).

The breakfast cereal industry provides an especially good illustration of the relationship between a lack of effective competition in a product market and excessive advertising expenditures. The industry is dominated by four major firms and had the second-highest expenditure on advertising as a percentage of sales of all of the more than three hundred categories of manufacturing industries in the United States. This industry was spending 18.5 per cent of sales on advertising in the late 1960s, compared to 3.8 per cent for consumer goods industries as a whole (Scherer, 1988:208).

Scherer cites evidence from one product category – ready-to-eat corn flakes – that illustrates the relationship between competition in an industry and expenditures on advertising. He notes that when Kellogg had virtual dominance of corn flake sales, its advertising outlays averaged 16.5 per cent of sales for those products. But when another company, Ralston, began to compete for the corn flake business, Kellogg's expenditure on advertising relative to sales began to decline, and by the early 1970s, in a more competitive environment, its corn flake advertising had declined to 5.8 per cent of sales. As Scherer writes, "The more price-competitive the situation became, the less incentive there was to support high levels of advertising.... In this respect, much cereal advertising is seen to be a social waste: an expenditure that would not have occurred if price competition had been working" (1988:213).[9]

Although corporate concentration may be more acute in the breakfast cereal trade than elsewhere, the lack of competition there and the excessive advertising expenditures it promotes are by no means unique to this sector of the food industry, or to other key sectors of advanced capitalist societies, from automobiles or petroleum products to a host of other consumer-oriented industries. The rise of the mammoth selling apparatus in the modern food system is intimately linked with the dominance of the industry by fewer and fewer firms. One cannot exist without the other. And in the end, unless it can be firmly established that this sales effort does serve a vital socio-economic role, it must be viewed as a tremendous social cost to be borne by society as a whole, one of the costs that follow in the wake of excessive corporate concentration.

So potent is the ethos of the marketplace in our society that we very rarely question the diversion of these very considerable resources of society towards activities of such questionable merit. Perhaps this is because we have been told by academic economists or industry apologists that advertising serves an important function in providing consumers with important information necessary to make rational choices in the marketplace. You do not have to "consume" very much advertising to develop a healthy scepticism about these claims. A casual scrutiny of TV commercials for children's breakfast cereal or for beer reveals little in the way of content that informs rational choices. This is not to say that the advertising has no message, but only that the message has more to do with promoting *irrational* choices than with aiding consumers to make rational decisions. This contention is supported by research that has attempted a rigorous assessment of the information content of advertising, in particular a study by A. Resnick and B. Stern (1977) that examined the information content of food advertisements on television.

Resnick and Stern judged a commercial to be informative if it contained at least one of fourteen informational cues (for example, information on price or value, ingredients, taste, guarantees or warranties, safety, nutrition, or new ideas). Clearly, this is a lenient test of the industry's claim that advertising provides a useful function for society. But even applying criteria as lenient as this, Resnick and Stern found that only 46 per cent of television food commercials could be considered to be "informative." Only 13 per cent of the ads surveyed had as many as two informational cues, and only one out of 144 advertisements had as many as three cues (Resnick and Stern, 1977:52-53). The authors conclude: "An important implication stemming from the study's results is that non-informative advertising might be an implicit admission that the product so described *fails to fulfill any unique or relevant needs of the consumer – be they taste, value, etc.*" (ibid., 52, my emphasis).

The reality of today's food business is that the big players spend vast sums in an effort to create a successful illusion around their products, which in the end might result in a shift in their market share in a product category of one or two percentage points. This is the situation in the Canadian brewing industry, for example, where just two corporations, Molson and Labatt, manage to spend about $100 million annually to protect and advance the sales of a very few competing beer brands they own (*Financial Post*, Oct. 24, 1988) – brands that are virtually identical in their essential quality and taste. While expenditures of such magnitude, largely to manufacture images and illusions around products, may be the most rational behaviour *from the point of view of the corporations* in their efforts to enhance their competitive positions, still we must ask whether this expenditure can in any way be considered rational *from the point of view of society and most of the people in it.*

To provide perspective it might be instructive to imagine what $100 million could be employed to do if *another* type of economic logic were at play in our society. For instance, the annual operating budget of the Daily Bread Food Bank, the largest food bank in Canada, was $640,000 in 1991. This organization had grown to meet the needs of some 150,000 people a month by 1992.[10] The advertising budgets of Canada's two leading brewing companies alone, then, would support the operation of over 150 similar food-distribution operations, and this expenditure would make a major contribution towards alleviating hunger among the poorest of Canadians. Alternatively, $100 million would pay the annual operating costs to provide day-care to more than 15,000 children and thereby significantly help to meet the pressing demand for day-care facilities in this country.[11] It could provide the poorest Canadians with a large subsidy for public transit use for a year: public transit cost has been found to be the most frequently cited "extra expense" related to the use of a food bank. The advertising budget of the two major Canadian beer companies is the equivalent of giving over 250,000 low-income people a 50 per cent reduction in public transit for one year.[12] These advertising expenditures are not trivial amounts, and any realistic assessment of the often-touted "efficiency" and "rationality" of the food system within capitalist societies must come to terms with the tremendous social resources that are channelled towards totally unproductive ends.

Foreign Ownership and Community Decline

Today more than ever Canadians are being continuously told by a variety of powerful people in business, government, and media that foreign investment capital is essential for the recovery and growth of our domestic economy. Lacking the wherewithal to do the job ourselves, or so we are told, we must not do anything that would discourage foreign investors; indeed, we have to moderate our demands for gender wage equality, improved social assistance, a national day-care system, and so on, so we will not scare off foreign investors. These days, foreign investment is supposed to be an entirely positive phenomenon, and if we can only capture our share of it we will be better off.

The food system provides an excellent example of the impact of foreign investment on Canadian society. The OECD characterizes Canada as having a high degree of foreign penetration in its food industry, higher than any other advanced capitalist country (1979:273; see also Warnock, 1978:66-70). Indeed, the increasing domination of U.S. firms in this area was characterized by the Federal Task Force on Agriculture in the late 1960s as "a major public policy issue" (Canada, 1969:235). Foreign ownership was especially apparent in the canned and preserved fruit and vegetable subsector, and in the flour and cereals, biscuits, confectionary, and soft drink subsectors. Since the signing of the Free Trade Agreement with the United States in the late 1980s, a new wave of foreign penetration of our food economy has begun – all part of the contemporary phenomenon of "restructuring."

For quite some time, anyone who follows developments in the Canadian food system has known that foreign ownership has had mainly negative consequences for the amount of research and development work done in the sector. A 1971 report, for example, estimated that a mere 3 per cent of all establishments in the food and beverage industry in Canada had "technical laboratories" capable of research and development (cited in Warnock, 1978:67). The OECD's 1979 study on the scientific and technical capabilities of member countries in the food industry noted that when the ratio between total R&D expenditures by business firms in the food industry and value added was calculated, Canada had the second-lowest value, above only Italy in this respect. Expenditures on research and development were also noted to be declining year by year in Canada, a fact the OECD argued was "a matter of concern" in Canada because "food is one of the most important industries, particularly in terms of exports" (1979:401, 405).

If foreign ownership of our food economy has meant that the pace and direction of development of this activity are largely determined beyond our borders, its impact has not been limited to that factor alone. Both historically and in more recent times the foreign takeover of important sectors of our food economy has played a vital part in shaping our food system, and not always in a direction most Canadians would have chosen. The experience of the food-processing and farming complex of Prince Edward County in Ontario is a case in point. This county, which is almost entirely surrounded by the waters of Lake Ontario, has a benign climate and reasonably good soils that have made it historically one of Canada's most favoured fruit and vegetable growing regions. More significantly, it was the location of the beginnings of Canada's canning industry, and by World War II only one other Canadian county had more canning factories at work processing local produce. But within less than thirty years after 1955, this well-developed farming and food-processing complex had all but disappeared.

The first part of the story of this decline is directly related to one of the largest foreign takeovers ever to occur in Canada's food industry. This was the 1956 sale of Canadian Canners, the largest fruit and vegetable canning company in the British Commonwealth, to California Packing Corporation (Del Monte) of California. Shortly afterwards this U.S. canning giant decided on a rationalization program that would jeopardize much of the canning industry in Prince Edward County. Opting to consolidate its production in plants in the southwestern part of the province, Canadian Canners began shutting down its thirteen Prince Edward County factories. Local efforts to purchase and utilize equipment in the plants were entirely thwarted by the company (Baxter, 1969:2), which preferred to destroy the machinery instead.

The reasoning behind Canadian Canners' decision to close down its factories and thereby deal a punishing blow to the canning industry in Prince Edward County may never be fully understood. The firm was notoriously secretive in its operations, and no mention of this period of consolidation is made in the company's official corporate history.[13] Clearly, a host of factors more closely associated with the decision-making process of a multinational enterprise, including the desire to co-ordinate production with U.S. parent plants, necessarily came into play once the Del Monte takeover had occurred.

After the initial shock of the Canadian Canners pull-out in the late 1950s, a few local independent processors were left to carry on the industry in the county. By the 1970s, to stay competitive these firms were under intense pressure to invest in new processing technologies – including purchasing automatic peeling equipment in the case of tomatoes and making significant increases in the speed of canning lines generally. The firms that were able to do so were in a better position to compete with the dominant firms in the industry. To further enhance their prospects, county processors set up a purchasing consortium so that inputs such as cans could be purchased in volume at prices comparable to those paid by the major firms. Several local companies also attempted to enhance their competitiveness by diversifying into non-traditional products and by producing specialty packs. According to the owner of one of the most successful of the local processing companies, these various strategies had placed the surviving independent firms on a competitive footing vis-à-vis the majors in the industry by the late 1970s.[14]

In the 1980s the U.S.-owned processing giants in the food industry once again made their weight felt in Prince Edward County. This time the excessive corporate concentration that foreign ownership had helped to foster in Canada was at the root of a period of *predatory pricing* in the marketplace. The pricing forced the county's remaining independent processing plants, firms that were apparently modern and competitive, to suspend their operations.

Part of the responsibility for the final demise of the Prince Edward County processors clearly rests with the provincial government of the time. The late 1970s had witnessed what effectively amounted to the dumping of cheap tomato paste from offshore producers into the Canadian market. This had driven ten or twelve canners out of business in the province.[15] However, GATT negotiations in 1981 established what was hoped would be a more stable price for tomato paste for provincial processors. At the same time the provincial government announced a new program, ostensibly to aid all processors of concentrated tomato products in the province. The program made considerable funding available for companies to expand facilities, but much of the funding was allocated to the largest tomato processor in the country, H.J. Heinz. The Prince Edward County tomato processors who were looking to benefit from a newly stabilized price were to be victims of this government policy, which substantially expanded the production capacity of the major player in the industry. As one former processor said, "The upshot of all this was that they [the government] cut the legs out from under us – the price of the finished product went all to hell again."[16]

The problem went beyond the world of tomato paste, however, to whole canned tomatoes, and soon after other major vegetable commodities as well. The provincial government funding had helped Heinz establish a "pilot plant" for whole canned tomatoes, the mainstay of many small packers. According to one well-placed informant, Canadian Canners was in the midst of yet another takeover bid, this one by the Nabisco corporation, and Heinz moved in to take advantage of the company's vulnerability and expand its own share in the tomato products business.[17] Moreover, the informant told the author:

> Green Giant [Pillsbury] saw this as an opportunity to take a crack at them [Canadian Canners] in the corn and pea business and the green and wax bean business in cans. So all of a sudden, tomatoes, corn, peas, and beans are all under price pressure from H.J. Heinz and Green Giant.... All of a sudden we get forty-nine cent specials all over the place. At forty-nine cents you can't make any money.
>
> You have got a guy who's been in the business one hundred years, who does peas, beans, and tomatoes, all of a sudden all of his commodities are under extreme pressure from all sides, and he wonders what the hell happened.
>
> You have another guy who does peas, tomatoes, and pumpkin pie filling. All of a sudden he is up to his ass in alligators....
>
> They all threw in the towel. They didn't see the sense in losing a lot of money. They were forced by the wayside. They were the last of the independent vegetable processors.

An independent food processor who was one of the victims of this period of predatory pricing summed up his experience:

> Unfortunately, as far as we ourselves were concerned, Pillsbury and Del Monte and their parent companies went into battle in the marketplace, and forced prices down, for example in the year of 1984 prices went down some 23 per cent, and this continued for longer than we could hack it....
>
> We were hanging in on tomato products [under pressure by Heinz at

this time], but when they [the majors] took a whack at other products, such as green peas and green and wax beans, it put the crunch on us....

So I made the decision ... and after a lot of soul searching since the plant was established in 1912 and I had been at it since 1952, that my best move ... was to salvage what I could and get out.[18]

The years of predatory pricing in the vegetable processing sector, lasting roughly from 1984 to 1986, essentially broke the back of the processing industry in Prince Edward County. It was not a conspiracy of the dominant firms to drive them out. Rather, it was the consequence of corporate strategies designed to increase market share, as conducted by a few oligopolistic foreign-based food-processing corporations with enormous economic power in the marketplace. The provincial government's predilection to subsidize the expansion and modernization of the activities of the most powerful players in the game only hastened the pace of the more general process.

Foreign penetration of the Canadian food system has not, typically, served to increase the efficiency of the business by offering healthy competition to existing Canadian firms. Historically it has tended to escalate the degree of concentration existing in the industry, because it entailed the establishment of businesses that were considerably larger than the majority of the firms existing in the industry. In more recent times, foreign subsidiaries operating in Canada have been able to draw on the considerable resources available to them as part of multinational operations, using those resources to engage in predatory pricing policies oriented to expanding their share of the market. The smaller independent Canadian-owned firms without such deep pockets were pushed to the wayside. The net result was to further aggravate concentration in this sector of the economy.

The Role of Agribusiness in Restructuring Farming

Alongside the issues of social waste, inefficiency, and growing inequality there is one other facet of the capitalist sector of the Canadian food system that has definite consequences for agricultural producers and also, we believe, for the wider rural community they belong to. This is the growing role of modern agribusiness in co-ordinating the agricultural sphere to secure corporate needs such as regularity of supply, consistency of quality, and profitability. This aspect of the food system also points out the very considerable inequalities in power among the various players.

Recent decades have seen a trend towards a growing integration of agribusiness corporations – input suppliers and especially food processors –

and producers. As data for the United States indicates, even by 1960 agribusiness firms had turned away from open-market purchasing for a number
of raw agricultural commodities towards arrangements that maximized their
control over agriculture. Much – and in some cases virtually all – of the farm
output for commodities such as fluid milk, broiler chickens, processed vegetables, seed crops, and citrus fruits was produced either through some form of
contracting arrangement or on the corporate farms that input manufacturers
and food processors operated themselves. By 1980 the tendency was for even
more output to be "sourced" by such means (see Table 3). This development
poses several questions. What is pushing this integration in the first place?
What type of integration is most favoured? And what are the most salient consequences of this integration for primary producers?

Table 3
U.S. Farm Output Accounted for by Production Contracts and Vertical Integration, Selected Commodities, 1960, 1980

Commodity	1960	1980	% change
Production Contracts (%)			
Sugar beets	98	98	(0)
Fluid milk	95	95	(0)
Broiler chickens	93	89	(–4.3)
Processed vegetables	67	85	(26.8)
Seed crops	80	80	(0)
Citrus fruits	60	65	(8.3)
Turkeys	30	62	(106.6)
Eggs	5	52	(940.0)
Sugar cane	40	40	(0)
Other fruits/nuts	20	35	(75)
Total farm output	15	23	(53.3)
Vertical Integration (%)			
Sugar cane	60	60	(0)
Eggs	10	37	(270)
Fresh vegetables	25	35	(40)
Potatoes	30	35	(10)
Citrus fruits	20	35	(75)
Turkeys	4	28	(600)
Other fruits/nuts	15	25	(66.6)
Processed vegetables	8	15	(87.5)
Broiler chickens	5	10	(100)
Seed crops	0.3	10	(3,233.3)
Total farm output	4	7	(75)

Source: Calculated from Kenneth R. Krause, *Corporate Farming, 1969–1982,* Washington, DC: USDA, 1987, Table 9.

It appears that the shift of food processors away from open-market purchases of agricultural inputs is the result of pressures stemming from the development of the productive forces in the capitalist sector of the food system. A primary imperative has been the heightened pressure to purchase the main input – agricultural produce – at the lowest prices possible, especially given the pressures on small and medium processors to be price competitive as retail concentration progresses. In addition to low cost inputs, the advent of mass production and continuous processing technologies producing standardized products have forced food processors to secure supplies of a specified quality and uniformity as well as to find ways to lengthen the agricultural production cycle so they can more fully utilize expensive processing equipment.[19]

The safest solution for processing capital would seem to be vertical integration into farming, that is, for food manufacturers to produce the food they need themselves. Indeed, in the 1970s it appeared that corporate farming would become the dominant trend in agriculture, as many non-food corporations, especially in the United States, were attracted to large-scale farming by a variety of new economic incentives. This trend was short-lived, and soon afterwards capital began to retreat from this form of production as economic incentives changed (see Carlson, 1971; Corditz, 1978). However, corporate farming by food-processing firms did not cease to exist, and by the late 1980s it was still significant in a number of commodity sectors in the United States (see Krause, 1987).

According to one of the few reports in existence dealing with corporate farming in Canada, processor integration into farming had become a matter of considerable controversy by the early 1960s because of the many vegetable farmers it displaced (OMAF, 1972:51). Vertical integration was reportedly stimulated by the availability of considerable land being held by speculators, who would rent it cheaply to processors. The development of new but expensive mechanical harvesters for some vegetable crops at that time was also credited with giving vertically integrated processors an advantage over smaller independent growers. Acreage controlled by processors for such key crops as green and wax beans, sweet corn, and green peas expanded at the rate of 74 · per cent, 43 per cent, and 33 per cent respectively between 1960 and 1970 (OMAF, 1972:50). Moreover, my own more recent survey of food processing and farming in the fruit and vegetable sector in Ontario showed that vertical integration into farming is not a thing of the past. About one-half of all firms surveyed had corporate farming operations and, as might be expected, the larger the firm, the larger the average size of its corporate farm (Winson, 1990).

In Ontario, at least, processors engaged in corporate farming were typically those who processed a combination of corn, beans, and/or peas, whereas firms processing vegetable crops such as tomatoes, cucumbers, and peppers or fruit crops such as cherries, apples, and peaches were much less likely to vertically integrate into agriculture. Thus, the characteristics of the agricultural commodity, rather than the size of firm, appear to be the more significant determinant of whether firms will be vertically integrated or not. Size of firm, as expected, does influence the average amount of land that vertically integrated operations will control. As to why agribusiness firms choose to maintain corporate farms in the case of commodities such as corn, peas, and beans, a few key factors appear to be critical. Firstly, harvesting equipment for these crops is expensive (about $180,000 in 1987), and few small farmers are willing to make an investment of this magnitude in equipment, given the modest returns associated with the crops. Therefore processors essentially have to use their own equipment to harvest the crops of independent farm operators. It thus makes sense for processors to try to more fully utilize their equipment and the labour force needed to operate it by having their own farming operations. As one multiproduct processor remarked, corporate farming "keeps equipment in use, and keeps the crew employed more of the time."[20]

Related to this as well is the high degree of *perishability* of these crops, and at least one processor noted that it was important to have the control and security of supply that corporate farming offered. Moreover, the country's largest corn processor indicated that his firm was heavily integrated into farming because it was very concerned with the use of the special hybrid variety of corn that the firm had developed. Management of this firm felt more secure working only with the few larger growers they had dealt with for some time and felt they could trust, while sourcing the rest of their supply from their own farms.

Despite a notable tendency by processors to favour vertical integration for some commodities, it was rare to find a firm that secured all its supply from its own operations. Why was this the case, given the factors that encouraged processors to go into corporate farming in the first place? Two main reasons came to light during the course of interviews. In the first place, the prices of the commodities fluctuate from year to year, although the presence of marketing boards with negotiating powers over price tend to reduce the ups and downs as compared to an open-market situation. More than one processor noted that when negotiated prices were lower they felt it was cheaper for them to obtain supplies from family farm operators than from their own corporate farms. Another factor processors mentioned consistently was the desire to minimize the risks due to weather. This could best be done by procuring produce from a number of farmers dispersed over as wide a geographic area as practicable, something that is difficult to do by relying on corporate farming alone.

Does corporate farming represent the main trend in agribusiness integration, and will it therefore come to affect the viability of the population of family farm operations that currently supply food processors? My study indicated that although farmers in this sector were in an increasingly precarious situation, it was not because of the expansion of corporate farming, but rather more because of the threat of the Free Trade Agreement, which exposes Canadian fruit and vegetable producers to unfettered competition with U.S. producers who typically work with larger production units and lower costs for labour and other key inputs. Indeed, only one firm had increased its corporate farming acreage by any appreciable amount in recent years, and it had no plans for further expansion. The majority of firms had stabilized the size of their agricultural operations. A few were decreasing their acreage, including one firm that had previously had one of the largest acreages in the province (at 13,000 acres) and was now leaving corporate farming altogether. Several processors that had ceased corporate farming, or reduced their acreage, said they had come to the conclusion that it was better for a processor to concentrate on its primary activity (processing) and leave farming to independent farm operators.

In one case at least, that of a major potato processor, the unprofitability of its corporate farming operations had only been discovered by accident. The plant manager's discussion of this subject speaks volumes about the efficiency of multinational corporate bureaucracies. He noted that the company's farming operations and processing were run as two separate divisions of the larger corporation. This was changed four years earlier, when corporate farming was integrated with processing operations. "When we started going through the books and taking a close look at it, what we found was that they [the corporate farm] hadn't been losing money every year according to the books, because what they had been doing was charging a brokerage fee for supplying the plant. That brokerage fee varied according to how much they needed to make the operation profitable. We just didn't have the efficiency."

Processors indicated that they would only consider expanding their corporate farms if the marketing boards for some reason in the future started to approve prices above the level that processors considered "acceptable." Even this unlikely scenario would probably see processors attempting to import cheaper U.S. or European produce into Canada, rather than substantially expanding their own agricultural production. As the executive of the largest integrated fruit and vegetable processor noted, "If necessary, in some product lines we can go across the border, for example in tomatoes, and we can get peas from California, peaches from New York State if we have to, and beans from New York as well."[21] Only with the most perishable vegetables, principally beans and corn, did he foresee major problems in substituting imports for local production if prices got "out of line."

Contract Farming

In Canada, as elsewhere, there is evidence that food manufacturers have favoured more and more contractual linkages with primary producers, rather than with corporate farming, as the primary means of securing agricultural produce. Why has this been the case? One long-time observer noted: "First, agriculture is very capital-consuming – that is, it usually takes far more capital to grow or produce a given quantity of farm products than to process and distribute that same quantity. So most food firms will prefer to put their capital elsewhere. Second, most agricultural products are produced under very competitive conditions with small profit margins. Big corporations usually put their capital where it can earn the highest possible rate-of-return-on-investment" (A.C. Hoffman, quoted in OECD, 1979:150).

A further disadvantage of corporate farming is that it is forced to bear all the risks posed by natural phenomena – weather, disease – that were once borne by the farm operators themselves. These factors help to explain why processors in a wide variety of commodity sectors have come to favour some form of contractual arrangement with family farm operators as a means of obtaining the maximum possible technical guarantees on produce characteristics at the lowest price. Also, production contracting, which is also referred to as forward contracting or virtual integration, has, by freeing up capital for investment in more profitable spheres of production, been one of the central conditions allowing large food processors to undertake lucrative foreign investments and develop multinational operations in recent times (OECD, 1979:159-60).

Farmers may have their own reasons for favouring a contracting arrangement: factors such as the price instability of farm commodities, overcoming problems related to the perishability of some farm produce, and even pressure from creditors (see Davis, 1980:151). Production contract arrangements have varied by country, region, and commodity type. In Europe arrangements have varied from contracts to purchase – which merely specify the quantity, price, and delivery times – to contracts entailing complete control over production by the processor, including supply of inputs and responsibility for management decisions concerning the agricultural production process (OECD, 1979:152). At this extreme, the farmer is more and more reduced "to a wage earner and works under supervision to produce commodities which he does not own" (ibid., 142).

In the United States contract farming is especially prevalent in the production of sugar beets, seed crops, fluid milk, broiler chickens, and to a lesser extent processed vegetables (see Table 3). In Belgium and the Netherlands, in certain areas of agriculture not usually the domain of contractual arrangements, such as the production of calves for veal, the great majority of production occurs under contract (OECD, 1979:155).

Why has this type of linkage been favoured by processors, at least in the case of certain commodities? For processing firms, production contracting can provide the control they require over agricultural production (that is, over quantity and quality of raw product), with the minimal outlay of capital (see Davis 1980:144). This control entails influence not only over the production process, thereby maximizing productivity, but also over the off-farm exchange process, maximizing the transfer of surplus from farm operators to the capitalist sector. Moreover, for processors the effectiveness of contractual arrangements in securing these objectives turns upon such factors as (1) the ability of the contractor to determine the contract price, and (2) the degree to which a farmer's access to product markets is limited or restricted (ibid.).

What have been the consequences for primary producers of this form of integration into agribusiness operations? The limited research available tends to confirm a basic tendency: integration via some form of contractual linkage typically entails the *loss of producer autonomy and control* over basic production decisions on the farm (see Clement, 1983; Davis, 1980; Winson, 1988). A study of the food system in Europe had this to say about the dairy sector in France:

> Not only do small and medium sized producers no longer have a choice regarding the general structure of their farms and production methods, but they are increasingly losing their freedom to decide to whom they sell and on what terms. Producers often find it difficult to change their collecting firm, especially since processing firms are amplifying their network of links with suppliers by providing dairy production plant, animal health aid and technical advice, and even marketing calves.... In short, processing firms are gaining increasingly tight control over dairy production. (Quoted in OECD, 1979:157)

In a study of contract farming in the United States, John Davis argues that contract farming grants the corporate processor a degree of control over both the on-farm production process and the off-farm exchange process. Davis notes: "At its most extreme, it may reduce the farmer to a wage earner on his own land – a piece-worker who provides his own tools and works under supervision to produce commodities which he does not own. He sells his labour power instead of chickens, apples, beans, or beets" (1980:142).

A study in Quebec found that most contract integration there was initiated by the farm input industries, principally food manufacturers and suppliers, rather than by food processors. This contract integration had significant negative implications for farm operators, entailing a loss of autonomy and bargaining power around input purchases and output sales, and a loss of control over basic production decisions (see Mehri, 1984:46-53).

Case Study No.1: Farming and Food Processors in Nova Scotia

Capitalist processing firms have often shown a preference to secure their raw supplies via some form of contractual arrangement. The case of the farming and food-processing chain in Nova Scotia in the late 1980s illustrates the effect of these contractual linkages on the internal operation of a farm, and the differences that exist among producers of different commodities.[22] But the situation in Nova Scotia also indicates that processors secure the production of some commodities using detailed written contracts, while acquiring other commodities without the benefit of even the simplest form of purchasing contract.

The fruit and vegetable subsector illustrates both extremes. In the case of most processing vegetables – for example, peas, beans, carrots, beets, cauliflower, brussels sprouts – farmers are required to sign a written contract that specifies conditions such as:
(1) the price processors will receive;
(2) the acreage the farmers must plant;
(3) the standard of seedbed preparation;
(4) the varieties to be planted will be decided by the processor;
(5) the final decision over the application of chemicals to combat weeds and pests, and application of fertilizer will be made by the processor;
(6) the harvested crop must meet standards defined by the processor;
(7) the processor will provide an inspector to determine the quality of the harvest; and
(8) the grower will not contract with another processor.[23]

In the case of peas and beans, further restrictive conditions are specified, including:

(1) a minimum acreage is to be planted, as decided by the processor;

(2) the processor will have access to the grower's land at any time for the duration of the contract;

(3) the processor has the right to determine planting and harvesting time and is solely responsible for harvesting the grower's crop;

(4) the processor has the right not to harvest or pay for any portion of a crop that contains an "unreasonable" amount of weed, is located on land not convenient for the operation of machinery, and/or exceeds the grower's average yields in recent years by more than an amount specified in the contract; and

(5) the processor shall deduct harvesting costs from monies owed the grower, and the crop remains at the grower's risk during the harvesting operation.

In the case of potatoes, farmers producing under contract have had to shift to lower-yielding varieties favoured by the processors, but with no compensation for the decline in yields. Moreover, a processor's inspector graded the crop and had final authority in deciding on the suitability of the potatoes. The farmer was levied a fee for this "service."

Why would farmers choose contract farming at all given these rigid and restrictive conditions? The motivation depends somewhat upon the commodity in question, but in the end contract farming is the most lucrative choice, given the few options available. In the case of potato growers, the contract price is generally better than any other price that can be obtained in the chronically depressed fresh market. With crops such as beans and peas, prices are not especially attractive, but given the necessity of crop rotation it makes more sense to grow a vegetable crop under contract and receive at least some return than to leave land fallow and receive nothing.[24]

If the linkage between farmers and processors is highly structured in the case of vegetables for processing, the same cannot be said in the case of apples, which have historically been the most important field crop for Nova Scotian agriculture. For the production of apples that went for processing in the province, there were no contracts of any type. The relationship between processor and grower centred on a meeting held once during the year during which the largest processor announced its price for apples. As a research report of the Nova Scotia Fruit Growers Association noted, this meeting "is not a structure conducive to effective negotiation. Processors are reluctant to divulge their needs lest their competitors pick up too much information on marketing tactics. In recent years, at least, the meetings have tended to be very short and communication has tended to be one-way. The largest firm more or less announces what it is prepared to pay" (NSFGA, 1985:37). In other words, the influence of producers appeared to be minimal. The processors followed up at some point during the growing season by sending a letter to growers specifying the price they were willing to pay. Beyond this the processor's role was limited up to the point when the crop was delivered and had to be graded.

What of the other agricultural subsectors in which farm operators directly supply processing operations? Dairy farming, the most significant subsector in Nova Scotia in terms of value of production at the farm gate and number of producers, has for a long time incorporated a substantial degree of control over farmer/suppliers. At one time farmers contracted milk to processors who could set the price of milk and the volume they were willing to buy. By 1970, after years of chaotic market conditions and considerable political debate, a national supply management system was established whereby provincial marketing boards would act as agents for farmers, buying and selling their milk, while processors negotiated a price with their board on a yearly basis. In this subsector, processors tightly monitor on-farm production through field representatives and constant in-plant testing of the milk they receive. While dairy farms are not subject to the intrusive contractual conditions found in the vegetable subsector, nevertheless dairy processors play a major role in determining such matters as technological change on farms. Such on-farm innovations as bulk-milk storage technology have had major negative consequences for smaller dairy farmers in particular.

What, then, accounts for the wide divergence in linkages between farmer and processor among the different commodities studied? The Nova Scotia study reinforced the conclusion that the *nature of the commodity* itself helps determine whether processors seek to establish a considerable degree of control – typically through a highly structured contractual arrangement – over the production of the commodity. Vegetables in general, and especially beans and peas, are highly perishable commodities, and their quality can be markedly harmed by improper horticultural practices and delays in harvesting and shipping. A representative of a large fruit and vegetable processing firm noted, for instance, that in the case of peas quality can be impaired by a harvesting delay of as little as a day, or even a period of hours. Apples, by comparison, are not nearly so perishable and can be stored, under appropriate conditions, for several months after they are picked and before they are finally processed.

Fluid milk is another commodity distinguished by its perishability and susceptibility to contamination. To this must be added the imperative imposed by modern processing technology, which involves the processing of large quantities of the raw material on a continuous basis. This imperative applies to both capitalist and co-operative processing operations. Strict quality control at the farm level can be viewed as a vital precondition for a modern dairy operation to produce highly standardized products in large volumes. Much the same can be said for the poultry subsector. The introduction of labour-displacing technology such as modern cut-up machinery requires the production of a "standardized" bird of a specific size and weight. To this is added the further pressure for standardization emanating from such high-volume buyers as the fast-food chains, which have rigid size and weight specifications that processors must meet to maintain their business.[25]

While the evidence supports the conclusion that processors have a preference for a highly structured linkage with farmer/suppliers if the commodity in question is especially perishable, and given certain imperatives imposed by processing technology itself, these factors alone did not entirely determine the content of all the contractual relationships studied. The wider political economy also has an influence on the makeup of the farming-processing chain. This includes the degree of corporate concentration at the processing end and the presence or absence of such quasi-state entities as product marketing boards, which have the power to decide on the key questions of price and contract conditions. Another factor that can be of considerable significance is the presence of a substantial co-operative processing sector to serve the needs of farmers.

Corporate Concentration and Primary Producers

From the point of view of individual processors, corporate consolidation is a necessary strategy to ensure a position in the marketplace, given current trends in the processing and retailing fields. From the point of view of farm operators, consolidation means fewer buyers for their production. This has indeed become the reality in Nova Scotia for farmers supplying processing firms. Until the 1980s, farmers growing vegetables for processing had at least two major buyers for their produce. By 1983, a situation of *monopsony* existed in this subsector, as the two remaining firms were consolidated into one corporate entity, leaving only one buyer in the market. Moreover, this firm later expanded into Ontario as well, so that it no longer had the same dependency on Nova Scotian produce. Should this firm decide to shift its operations out of the province at some future date, growing vegetables for processing in the province would no longer be a viable proposition.[26] Farmers who produced potatoes for processing in the province faced the same monopsonistic conditions. The potato farmers grow a particular variety of potato for processing purposes and cannot readily turn to the fresh market with their crop. Prices in the fresh market have been low and generally below those available from the processor, diminishing the attractiveness of this option for potato farmers.[27]

The situation for apple producers, long the mainstay of agriculture in the Annapolis Valley, is affected by the same process of consolidation. Of the three firms still operating in 1987, one, a co-operative organization, had not re-established its own processing operations since fire destroyed its facilities several years earlier. One of the two private apple processors was a small operation, producing only juice. The presence of this firm could in future be the only factor preventing the formation of a monopsony in this subsector as well. If it disappeared there would be little resistance left to pressure from the large remaining processor to force prices to growers substantially downward. Indeed, the state of competition in apple processing in Nova Scotia has been recognized as a matter of concern by the province's apple producers (NSFGA, 1985:37).

In the other main commodity subsectors – dairy and poultry – high degrees of buyer concentration are also the norm, especially given the consolidation of the once numerous smaller dairies to a point where three firms dominate the provincial market. Moreover, dairy farmers in certain parts of the province effectively have access to only one dairy. In this subsector, and that of poultry, the powerful presence of co-operative processing firms may modify the effect of a high degree of corporate concentration, as far as farmers are concerned.

The Significance of Product Marketing Boards

Farmer/processor relationships can also be modified by the existence of product marketing boards, which have typically been the outcome of producer initiative and political pressure (Matthie, 1982:6). In promoting marketing boards, producers have aimed to (1) raise the price of farm products; (2) stabilize these prices; and (3) improve the bargaining position of farmers vis-à-vis the relatively more concentrated processing and retail sectors (Prescott, 1981:8). The degree of a marketing board's influence, however, is determined by the powers it has over negotiating such matters as contract conditions and price, powers that are established by provincial or federal legislation.

In considering the impact of a product marketing board it is useful to contrast the situation in Nova Scotia with that of Ontario and to look at a subsector in which considerable interprovincial variation exists, such as processed vegetable production. Boards in this subsector in Ontario have the power to negotiate price and contract conditions, whereas in Nova Scotia the boards serve primarily to provide producers with information and have no real powers to determine contract prices and conditions.

With respect to contract conditions, several important differences characterize contracts negotiated in Ontario for producers with the aid of the Ontario Vegetable Growers' Marketing Board, in comparison with contracts prevailing in Nova Scotia. For example, agreements governing contracts in Ontario typically specify that a third party, usually a government inspector, be involved in the grading of the grower's crop and that the costs of such service be shared equally between grower and processor. The agreements also specify that growers can request regrading of their crops and, where possible, that grading be made with the aid of standardized instruments designed to minimize the role of "human error" in this process.[28] Other provisions include the requirement that processors pay farmers/suppliers the full amount of the purchase price of their crops within a few weeks after delivery of their harvests; that where processors provide services for insect control the cost of these services be shared equally by growers and processors, rather than be covered by growers alone; and that in cases where processors provide seed to growers, the processors are required to supply seed that will provide a specified minimum standard of germination. Finally, in Ontario growers whose crop is not harvested for reasons beyond their control are liable for a much lower percent of the crop potential than is the case in Nova Scotia, where no body exists to negotiate such conditions on behalf of growers.[29]

The comparison of contract conditions in Ontario and Nova Scotia suggests, firstly, that regardless of the presence or absence of a mediating body such as a product marketing board, processors have a high degree of involvement in crops that are distinguished by their perishability, and a lower degree of involvement in the production of agricultural commodities that stand up better to the rigours of harvesting, transport, and storage. Secondly, the presence of a product marketing board with legislated powers over contract conditions can result in contracts that, in matters such as provision of services by the processor, grading of the crop, grower liability when the crop is not harvested, and terms of payment, provide an important degree of protection to the grower in cases in which processor abuse is possible because of their much greater economic power.

The Question of Price

One further and most essential matter over which a marketing board can exercise its power is price. In the case of vegetables produced for processing in

Ontario, the Marketing Board has the power to negotiate price. Where a price acceptable to producers and processors cannot be negotiated, provision is made to pass the matter to an independent arbitrator with final decision-making power (Prescott, 1981:32-34). Can we expect this to lead to more favourable prices for producers than in a situation where no such board exists, as in Nova Scotia? Presumably this matter also turns on the degree of corporate concentration characterizing regional processing sectors. Certainly, if a high degree of concentration is present where no marketing board exists to negotiate price, the farmers/suppliers, who are typically small economic actors and poorly organized, have very little leverage in the marketplace. The data on prices that producers received for several vegetable crops in Ontario and Nova Scotia in 1985 would tend to support this conclusion (see Table 4). Data from a marketing study completed in the late 1960s indicate that a price differential has existed for at least twenty years for crops such as peas, beans, and potatoes, and that it also existed for apples (Warnock Hersey, 1970:37-41). Indeed, prices to farmers/suppliers in Nova Scotia relative to Ontario appear to have deteriorated substantially over this period.[30] There is, moreover, little reason to think that input prices would be any lower for farmers in Nova Scotia; in fact, quite the opposite is likely to be the case with respect to such inputs as feed, fertilizer, machinery, and ag-chemicals.

The evidence suggests that, as in the case of contract conditions, the presence of a marketing board with the power to intervene in the marketplace will assist producers in getting more attractive prices for their products. Indeed, this would seem to have been the case with other sectors of Canadian agriculture, such as the broiler chicken and egg sectors (Arcus, 1981:61). This, in turn, could be expected to put some pressure on the profits of processors, especially as they meet increasing competition from firms that are not affected by producer-controlled marketing organizations.[31] Interestingly, evidence from the Canadian broiler chicken and egg sectors suggests that although the margins of processors may be squeezed by marketing boards, those of input industries such as hatcheries and feed companies may actually be enhanced due to the improved financial position of farm operators who constitute the primary customers for these latter firms (see ibid.).

Dependent or Degenerated Commodity Producers?

The study of farmer and food-processor linkages in Nova Scotia illustrates that the production of highly perishable commodities tends to be governed by highly structured production contracts, especially if on-farm production is not

Table 4
Prices for Processed Vegetables, Ontario and Nova Scotia, 1985

Commodity	Price	
	Ontario	Nova Scotia
Green & wax beans	$234.00/ton	$185–200/ton[1]
Peas	$286.60–549.40/ton[2]	$300/ton[3]
Carrots	$68.85–110.60/ton[4]	$55.00–65.00/ton[5]
Cauliflower	$261.70–313.00/ton[6]	$520/ton
Potatoes	$5.66–9.35/cwt[7]	$5.91/cwt

Notes
1. The higher price was paid for wax beans. The average price received by bean growers in 1985 was actually closer to $185.
2. Price reflects quality as measured by tenderometer readings.
3. This is the average price paid for peas that went for processing.
4. Figures indicate the range of prices for that year, with the highest price being for the June harvest and the lowest price for the harvests in the September to November period.
5. Lower price is for carrots 1½ inch in diameter and higher price for carrots ¾ to 1½ inch in diameter. The average price paid to carrot growers was reported to be closer to the lower figure.
6. Higher price is for 'tied' cauliflower. The notably high price for Nova Scotia is not easy to account for. The processor may have been willing to pay extra to secure the small quantities that were needed, rather than truck a small amount of the raw product in from out of province.
7. Highest price paid in July. Prices declined each month until October and then increased in each of the following months. Most potatoes were sold in September at a price close to the best price of $5.66/cwt.

Sources: Ontario Vegetable Growers' Marketing Board, *42nd Annual Report,* 1986; 1985 growers' contracts from Nova Scotia, and interviews with staff of the N.S. Dept. of Agriculture, 1986.

also being regulated in some fashion by an external body – such as a marketing board as in the case of fluid milk. In the case of these commodities, many significant management decisions have been taken over by the capitalist processing sector, as well as such key production activities as harvesting. The "rights" of capital as detailed in production contracts with growers would seem to impinge even on the formal rights of ownership usually enjoyed by petty producers (such as right of access to the farm). However, in the case of some less perishable commodities – apples being the primary example – processing capital is much less intrusive, and growers conform more to the role of classical independent commodity producers.

The specific impact on producers of contractual forms of integration with food processors or input suppliers has led observers to argue for new ways to conceptualize what we have called independent or petty commodity producers. Drawing on studies of the social organization of both agriculture and the fisheries, Canadian sociologist Wallace Clement has argued that the thrust of capitalism is to proletarianize formerly independent producers, but not necessarily by transforming them into wage labourers. Rather, "The *form* of the petite bourgeoisie may remain while the *content* (in the sense of economic ownership) may be captured by capital.... A producer experiencing proletarianization may retain direction of his own labour power ... while the means by which capital is accumulated, the disposal of the products of labour, and the technical development of the labour process (economic ownership) may be dominated by capital" (1984:7). Thus, in sectors of Canadian agriculture and the fishing industry, "independent" commodity producers have become what Clement calls "dependent" commodity producers in certain contexts, that is, where direct producers are "compelled into a contract or monopoly relationship with capital," rather than selling on the open market. Ownership remains, but capital essentially co-ordinates the labour process. "It becomes social labour organized by capital" (ibid., 8).

Others such as John E. Davis (1980) have preferred to use the term "propertied labourer" to describe agriculturalists who have experienced this loss of autonomy and control. Elsewhere I have suggested the concept of "degenerated commodity production" because it more accurately describes the process involving a relative decline in producers' independence from external economic actors (Winson, 1988). Whatever concept is used, the point is that in these sectors of farming (and fishing for that matter) that have of necessity become involved in contractual linkages with processing firms, producers have generally lost something in the bargain. That something can be reduced to the notion of "independence." The degree of the loss varies with the type of commodity produced, and to some extent with the ability of producers to protect themselves through some form of collective organization or institutional apparatus elaborated to safeguard their collective interests. We shall return to this issue below.

Case Study No. 2: Agribusiness and Farm Differentiation in Ontario

If integration with agribusiness can dramatically alter the internal organization of farming, it seems logical to expect that it would also have some impact on the actual viability of farming operations – that is, upon the size and kinds of farms that survive and expand, and the ones that will be extinguished. Indeed, my research suggests that there is an impact on the farming structure, that the realities of integration favour some farmers and discriminate against others – a process called the *social differentiation* of the farm structure. Once again, as well, the concentration of economic power in the agribusiness sector has an important role to play in determining certain outcomes.

My findings on this issue are based primarily on a study of fruit and vegetable processors in Canada's foremost fruit and vegetable growing province, Ontario, in the late 1980s, although there is evidence from the dairy sector in Nova Scotia as well.[32] For some time in Ontario there had been many more farm operators seeking fruit and vegetable contracts than there were contracts available. This was in part related to the depressed prices farmers were receiving in the late 1980s for traditional alternative crops such as soybeans and wheat. It is the manufacturers that were deciding who would receive contracts in the first place, and the situation in 1987 was giving processors considerable discretionary powers in determining what kinds of farm operators would be favoured. Processors were therefore in a position to influence the type of farm structure by the way they rewarded contracts.

In the study we asked processors about their preferences with respect to the size of farm operations they would rather deal with. We also asked them to elaborate on the reasons underlying their stated preferences. Most processors had a definite preference regarding farm size and articulated a fairly clear-cut rationale to justify this preference.

With respect to both the large-size and medium-size companies the responses followed these patterns: (1) a definite preference for the larger, heavily capitalized farm operators; or (2) where a definite preference for large growers was not stated, a clear indication that the processor no longer wished to deal with "small" farm operators (see Table 5). There were only two exceptions to these two patterns, and in one case the firm in question was a large cucumber processor. For mainly technical reasons it had not been possible to mechanize the harvesting of cucumbers and still obtain the small cucumber sizes most desired by the processor. Thus harvesting continued to be a completely "hand-pick" operation, which provided better conditions for the survival of small-farm operators. A switch to obtaining the preferred smaller cucumbers by mechanical means would most likely alter the processors' preferences considerably, in favour of the larger growers.[33] None of the small processors, on the other hand, indicated a preference for the larger growers. In fact, with this group of processing firms, either a clear willingness to contract with small growers was indicated, or processors stated that they had no clear preferences with regards to grower size. Table 5 indicates a positive relationship between processor size and the size of the farm operation preferred by processors of a given size.[34] In the study I was also able to corroborate this in-

Table 5
Processor Preferences for Various-size Farm Operations,
by Size of Processor, 1987 (N = 20)

	Type of preference		
Firm	Type A[1]	Type B[2]	Type C[3]
Large[4]			
A1	X		
A2	X		
A3		X	
A4		X	
Medium			
B1	X		
B2		X	
B3	X		
B4			X
B5	X		
B6		X	
B8		X	
B9		X	
B10			X
B11		X	
Small			
C1			X
C4			X
C7			X
C10			X
C11			X

Gamma = .857, Somer's d = .654.

Notes
1. Type A refers to a definite stated preference for the larger, more capitalized growers.
2. Type B refers to a response where processors indicated that they were no longer interested in dealing with the smaller growers, although a definite preference for larger growers was not indicated.
3. Type C refers to a response where either a preference for small growers was given, or no preference concerning the preferred size of grower was indicated.
4. Firms were classified according to estimates provided by informants of the replacement value of all equipment and buildings associated with their processing activities. Large = replacement value > $50,000,000; medium = replacement value between $10,000,000 and $49,000,000; small = replacement value < $10,000,000. There are various ways one could attempt to estimate the value of the fixed capital of these firms. Our main objective, however, was to gain some idea of the *relative* size of each firm.

Source: Interviews with fruit and vegetable processors, Ontario, 1987, 1988.

formation on processors' preferences with actual data on acres contracted with each farmer for the different-size firms in the tomato subsector, the most important of the fruit and vegetable crops.[35]

Why did most medium and large processors have an aversion to contracting with small-farm operators? The rationale most frequently given was that large farmers were more efficient farmers, or "businessmen," and tended to turn out better-quality produce.[36] Several other rationales surfaced almost as frequently in the course of interviews. In summary they are (in order of declining frequency):

(1) processors favoured mechanical harvesting of crops, and only large growers had the acreage to justify this type of harvesting;

(2) administrative costs could be lowered by dealing with fewer but larger farms;

(3) large growers provided services that smaller growers could not afford (for example, trucks to haul produce to the processor or storage facilities for raw produce when a processor's warehouse capacity is insufficient);

(4) it was more convenient for the processor to harvest field crops from a few large farms than many smaller farms of dispersed acreage.

Some of these rationales provide clues as to why the small-processor sector is more open to dealing with small farms. First of all, most small manufacturers were processing produce from farms where mechanized harvesters were not used to any great degree, as with tree crops such as cherries and apples or a tomato variety used for whole-pack tomatoes that was not suitable for mechanical harvesting. Secondly, because small processors rarely processed crops such as corn, green beans, and peas where they themselves would organize the harvesting of the crop, they did not consider fewer farmers with larger acreage to be a great convenience. The owner of one small firm noted that by contracting with farmers using hand-harvesting methods, he was able to stretch out the processing season longer than a mechanized harvest would allow, which made for more efficient use of the processing equipment. Finally, small processors mentioned that it would be unwise for them to concentrate their source of supply in a limited geographical area by dealing with a few large growers. Having a larger number of smaller farmers provided geographic dispersion and minimized the risk of going without supply because of natural calamities, such as excessive rain or hail.

There is a definite relationship, then, between the size of the processor and the size and character of the farm operator that a processor contracts for production. As for the wider significance of this situation, I believe that the relationship must be viewed within the context of changes in the corporate structure of the food-processing sector itself.

In the late 1940s the number of fruit and vegetable processors in Ontario peaked at about 230 (see Figure 5). Since then the concentration of capital in this sphere has gathered momentum, forced by both pressures internal to the food-manufacturing sector and external pressures, in particular those created by the consolidation of the food-retailing sector. In Canada, supermarket food chains had only 23 per cent of food sales in 1950, but had expanded this to over 70 per cent by 1987 (Matas, 1987:1). Moreover, levels of corporate concentration in this sector are substantially higher than in the United States (Warnock, 1978:203). According to one tomato processor, "If you take a look at the power of the retail sector, in the United States it would have been stopped long ago.... When I think that when we started in 1945 we had over four hundred different customers, now six people have, what, say 90 per cent of our business." Another food-processing executive, this time representing one of the four largest firms in the country, noted about retailer power that there are "only five buyers in Canada, that's all there are. It's rather difficult to negotiate when you've only got five customers." [37]

The outcome of these various processes has been the differentiation of the food-manufacturing sector between, on the one hand, a small group of very large diversified corporations and, on the other, a more numerous group of

Figure 5
Number of Fruit and Vegetable Processors in Ontario, 1940–87

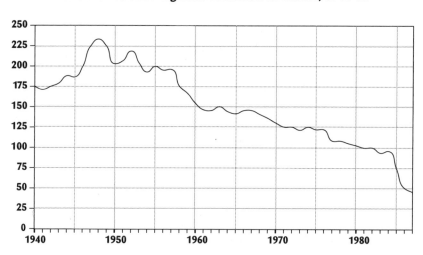

Sources: Dominion Bureau of Statistics, 'Fruit and Vegetable Canners and Preservers'; Statistics Canada, 'Fruit and Vegetable Processing Industries'; Ontario Food Processors Association, *Directory*, 1986–87.

medium and small firms largely subordinated to the dictates of the largest manufacturers they are co-packing for, or of the retail food oligopolies they supply under the manufacturers' private labels. The successive merger waves, the predatory pricing tactics of the largest firms, and the periodic impact of low-priced imports had battered Ontario's canning sector. One of the region's smaller tomato processors was especially candid about the impact of the predatory pricing practices of the largest firms on the small food processors:

> Last year ... one large firm, an international firm, decided they were going to expand their market portion of whole-pack tomatoes. They decided to do that by putting almost a loss-leader deal on tomatoes, so in the consequence the whole group of us suffered because the other large one that had a large portion of the Canadian market said, hey, this new big guy is not going to take a portion of the market we already had. Sure, they hurt each other, but these guys have the wherewithal to hurt each other, but the others are killed!
>
> We've had two bad years ... it wasn't a big pack last year, it should have gone at a fair price and I bet if you talk to most of the independents, until recently they would have lost money on whole tomatoes.[38]

By the late 1980s firms with plant and equipment valued at under $1 million had all but disappeared, and the total number of plants had declined to less than fifty by the mid-1980s.

This context of corporate restructuring, and in particular the decline of the small processors, is connected to the question of the relationship between processor size and preference for farms of a specific size. These contemporary corporate trends can only mean that the smaller farm operators will be ever more hard-pressed to secure contracts for their production in future. One of the processors interviewed summed up the predicament for small farmers and small processors alike: "I think the big ones [processors] are looking at going with only big growers, because they don't have the time to bring the small ones in.... So I think that the small grower is bound up with the small plants and, unfortunately, when he looks at the small plants his risk is greater than the big growers, for the simple reason that a small plant can go broke quicker than a big plant."[39]

The evidence suggests that corporate food processors have also fostered the social differentiation of farms through their practices in awarding production contracts. The larger processors have tended to award contracts to the larger, more capitalized growers for a variety of reasons, ranging from administrative convenience to the perception that large growers have greater capacity to provide quality produce consistently and to provide storage and haulage services to the processor. Furthermore, it is the power of processors to award much sought-after production contracts that makes this bias towards the large farmer operative.

It is the *process of corporate concentration* that really determines the significance of this bias on the part of the larger processors. My investigation of the fruit and vegetable sector provides evidence that small processors are essentially the refuge for the smaller farm operators. These small processors have borne the brunt of corporate concentration in this sector of the economy. Given the absence of attractive alternatives to contract farming, the demise of the small processors is undoubtedly a critical factor in the demise of the small growers as well.

The Role of Technological Change

There is one other way in which the close integration with food manufacturers has had negative consequences for the smaller farm operations, and this is related to technological change on farms as promoted by the processors. Farms involved in a contractual form of linkage have been affected, but so too have farmers with other types of arrangements with processors.

One example of this is the situation of many producers of tomatoes that go for processing, a situation typically governed by a contract. In the case of Ontario, for example, the province's largest tomato processor has favoured farm operators who were able to mechanize their operations with a mechanical tomato harvester. Before tomato farms were mechanized, farmers could easily switch production to other types of crops – such as soybeans or wheat – in years when tomato prices turned out to be unattractive. This flexibility was a disadvantage for processors, who wanted to be sure of a certain volume of supply. Having farmers purchase expensive harvesting equipment – equipment good for harvesting tomatoes only – has served to lock in farmers to the continuous production of tomatoes and reduced their options for producing alternative crops. They cannot afford to leave such expensive equipment idle for a year. The purchase of a harvester also requires a certain minimum acreage for it to be economical, and this average is high enough to exclude the smaller tomato growers from being able to meet the processors' criteria and thereby receive a contract.

The processors' influence in promoting technological change on farms can have negative consequences even when the standard obligations and constraints of contract farming do not exist. The introduction of bulk milk-storage tanks on dairy farms is a case in point. The research in Nova Scotia illustrated this clearly. By the early 1960s or so, dairies there had decided that the quality of their end products would improve and their milk collection costs would decrease if farmers switched from shipping milk in cans to storing milk in bulk on the farm. The stored milk would later be picked up by tanker truck. The dairies all denied that they had ever forced farmers to switch to the expensive bulk storage equipment, but the manager of the province's second-largest dairy was more candid than most. He noted that the situation around the introduction of this innovation was particularly problematic.

> This was a bad one. For two years we had both [cans and bulk storage] and gradually more and more producers got involved as bulk shippers, and actually the quality was a lot better, so I guess then we announced that we would not take milk in cans in our dairy any more, and that anyone shipping in cans would have to take it to our creameries, where it would be used for ice cream and butter [and receive the lower industrial price]. We wouldn't pick it up any more, they had to take it in. We never wrote a letter saying that as of a certain date you would have to ship bulk, [but we] made it pretty damn inconvenient if they didn't.[40]

The social impact of this technological change impelled by processors was especially significant. As one government agricultural representative with many years of experience with local farmers remarked, "Bringing in bulk storage [of milk] must have cleared out half the dairy farmers in the province."[41]

The issue of technological change on farms illustrates another of the negative consequences of excessive corporate concentration. When farmers have few if any choices about whom they will sell their production to, they are forced to adopt the processors' preferences regarding on-farm technology. Those preferences have frequently involved technology that has seriously prejudiced the smaller farm operators.

The Farm-Community Nexus

A sociological analysis of the impact of agribusiness operations cannot stop at the farm gate. Fortunately there is a relatively rich tradition considering the farm-community nexus within rural sociology. The impacts we have discussed above need to be viewed in light of the research findings of this area, beginning with Walter Goldschmidt's important classic work conducted in the 1940s (1947, 1978). Goldschmidt argued that the change towards a small number of large-scale farms with industrial workplace structures produces social conditions that reduce the economic and social vitality of rural communities – a contention that has been occasionally challenged or more frequently modified in light of new evidence (see Harris and Gilbert, 1982; Green 1985; Korsching and Stofferahn, 1986; Lobao, 1987, 1990). Nevertheless, according to Louis Swanson numerous studies over four decades have demonstrated that: "Where there are few off-farm opportunities and family farming is the dominant source of community economic well-being, a decline in farm numbers and an increase in farm size will contribute to a decline in indicators of community well-being" (1988:8). Similarly, the processes examined in our case studies may well affect the structure of rural communities in the regions under investigation, and in a negative fashion.

The increasing pressure placed on small farmers by arrangements such as contract farming, in the context of the concentration of food manufacturers, is not, of course, the only factor to consider in examining the fate of rural communities. Other forces, such as the restructuring of non-farm enterprises – credit institutions, input suppliers, wholesalers, and food and non-food retailers – and the increasing geographic concentration of their operations (Mitchell, 1975; Raup, 1973) necessarily have an impact on towns and villages; and other factors that can modify the strength of the relationship discussed by Goldschmidt include the specific production and organizational features of local farms (Green, 1985:272), the mix of non-farm enterprises, and the level of community dependence on agricultural versus non-agricultural economic activities (Swanson, 1988:11). Nevertheless, in certain areas of the country rural communities depend substantially on the surrounding farm economy for their livelihood. For these communities, the high degree of integration involving agribusiness and farm operators is a preoccupying reality.

Concentration, Power, and Inequality

Whereas the direct producers of food had faced concentrated economic power largely outside the food economy in the early part of the century, in the form of the banks and the railway monopoly, after World War II they were forced more and more to deal with powerful corporate entities within the food system itself. Some sectors of the farm community were able to confront this threat through successful agitation for supply management legislation and producer controlled marketing boards. Other sectors were less successful and were left open to the full brunt of "market forces," which in the Canadian economy is generally a euphemism for a market controlled by a very few multinational conglomerate firms. However, even when producers did achieve some influence over their economic environment, it turned out to be only a partial influence, limited by the tremendous growth in the power of agro-industrial firms. In areas such as dairy farming and growing fruit and vegetables for processing industries, for instance, where producers have had some success in protecting their position through legislation, there has still been a continuing process of "rationalization" of farm operations, which is yet another euphemism, this time for eliminating the smaller operators. The most that can be said is that the survivors are somewhat better off financially than farmers in the crisis-ridden non-supply-managed sectors, and that the rate of attrition among farmers is somewhat slower for these farmers than for agriculture as a whole.

What about the other effects of the consolidation of the agro-food complex? The 1980s was a frenzied period of corporate merger mania and leveraged buy-out activity that has left the food business concentrated to an unprecedented degree. In the breakfast cereal industry, for instance, firms reached a size that had little to do with the maximizing of efficiency through economies of scale, but much to do with engrossing the bottom line via the maximum enhancement of economic power in the marketplace. Excessive market power in the hands of a few firms has real consequences: it can jeopardize the survival of efficient but smaller firms unable to fend off the predatory pricing practices of the conglomerate giants, and in the process it can undermine the farm operations that supply these firms – along with the rural communities they help to sustain. It can lead to chronically higher prices for food than consumers would otherwise have had to pay. And it can promote excessive expenditures on advertising and other promotional activities that ultimately result in a tremendous waste of society's scarce resources.

Why, then, have Canadians been willing to accept the status quo? Why have they not demanded government action to rectify this situation? One long-time observer of the Canadian corporate world offers the following provocative explanation.

Most Canadians suffer from strong feelings of insecurity. To gain security they have put their trust in large aggregations of power – both private and public.... Very few Canadians fear any potential loss of liberty associated with the concentration of power. Canadians have systematically traded off individual responsibility for powerful social and economic institutions they believe provide collective security.... Competition both as a social and economic process is little esteemed by

Canadians. Competition implies conflict and competitors can behave ruthlessly. Moreover, it is a disorderly process and Canadians place a high value on order and predictability. In general Canadians do not see a growth of aggregate concentration as a threat to either their personal political freedom or to their democratic political system. (Stanbury, 1988:430)

There may be more than a little truth to this view. Nevertheless, it is not the whole story. If Canadians do not see the present concentration of corporate power as a threat to the democratic process, perhaps we need to look to the mechanisms that in a democratic society are supposed to warn the public of such dangers. We need to ask, for instance, whether a "free press" that would normally serve this role can be said to exist in Canada, when the newspaper industry itself exemplifies a situation in which power is concentrated in extremely few hands. The other sectors of the Canadian mass media do not differ much in this regard, with the notable exception of public television and radio.

We also need to look at the ideology of the corporate world and recognize that this ideology is being promoted to the society at large by the mass media in a manner that is almost totally uncritical of the vested interests that lie behind it. We live in an era when bigness – at least with respect to the corporate world – is regarded as a virtue. It is bigness, we are told, that will allow Canadian firms to compete in the new global market. Any form of government intervention that thwarts corporations in their quest for bigness is portrayed as being against the general interests of society. In an important sense, though, this kind of argument is a basic admission that raw market power is the only game worth playing any more, that running an efficient operation is no longer enough to effectively "compete" in the marketplace. However, if a competitive market economy is supposed to be the most effective way of allocating the scarce resources of a society, and yet the only type of firm that can survive in the current market economy is the oligopolistic firm that by definition is anathema to competition, then we have to ask whether society as a whole is any longer being well served by organizing the economy along market lines. As we have seen, from the perspective of society, rather than from the perspective of the firm, ever higher levels of corporate concentration are very often not an efficient or a rational option.

It is also a problem that the social and moral implications of concentrating economic resources in so few hands rarely receive the wide discussion they deserve. This situation must be appreciated for what it is – a process by which the resources of society and the livelihoods of many of its citizens come under the control of fewer and fewer people. The corollary of this is that more and more people are dispossessed of their power to make effective decisions over their lives. To speak only of the food system, evidence abounds of a growing polarization that has only increased the social inequalities in the wider society. The evidence of this is all around us. By 1987, after-tax profits of food manufacturers totalled $1.26 billion. Removing the effects of inflation, this still added up to a 65 per cent increase from 1981 (*The Toronto Star*, June 26, 1988). These profits, it must be remembered, are being concentrated in fewer hands. At the same time the prices farmers received for their produce were down by an average of 10 per cent between 1981 and 1987, and for some major sectors of Canadian agriculture, such as grain farming, prices were down by over 50 per cent (ibid.).

The current inequities in our food system may be more starkly illustrated. We have now reached the point of concentrated economic wealth in the food system, whereby one firm can afford to reward its chief executive officer with a compensation package worth $89 million, on top of his regular $5 million salary, as the Atlanta-based Coca-Cola Company did in 1991 (*The Globe and Mail*, April 16, 1992:B7). But while a few food company executives are profiting extremely handsomely from their control over our food economy, unprecedented numbers of people in Canada – a wealthy country by all accounts – are simply unable to get enough food through what most of us would regard as normal channels. The rapid growth in Canadian food banks, from 126 in 1988 to 292 in 1991, shows the other face of the growing social inequality in our society. It is not just that a few people are getting very rich. It is also that many people are becoming very poor.

Although a number of factors are behind this trend, certainly one factor is that despite the general stagnation in farm commodity prices over the 1980s, food retail prices continued to escalate during this decade – by some 32 per cent between 1981 and 1987 alone (*The Toronto Star*, June 26, 1988).[42] As a major study on farm, wholesale, and retail price levels for agricultural commodities reported in 1992, much of the price increase for such basic foods as milk, cheese, chicken, and turkey comes *after* the processing companies have sold the product to retail food chains. In other words, the price increase is accounted for by the mark-ups of the retail food chains themselves. This data is a sign that other influential forces are also at play in our food system. These other forces are the retail food giants, and it is to their role in the Canadian food economy that we now turn.

CHAPTER SEVEN

FOOD RETAILERS: THE NEW
MASTERS OF THE FOOD
SYSTEM

● ● ● ● ● ● ● ●

*The retailers charge the manufacturers to bring a new
product into their stores, charge again for location, and
space on their shelves and then charge once more for a
prominent display in their advertisements and flyers.
With almost insatiable appetites, many retailers then
ask for additional payments over and above those
already provided.*

The Globe and Mail, February 28, 1987

More and more influence in the Canadian food economy has been captured by the complex of activities called "agribusiness." The great rural to
urban shift in our society over the last fifty years – the separation of the people from the land and a decline in self-sufficiency in food production and
preparation – has opened the door to an amazingly quick and then sustained
growth in demand for time-saving processed foodstuffs. The output of the
fruit and vegetables preparation industry, for example, more than doubled
from 1923 to 1929 alone (Canada, 1937:213).

In Canada activity in the food-manufacturing sector took two primary
forms. One was the limited liability company responsible to its owners –
stockholders in the case of a public company. The other form was the co-operative enterprise, which was, at least in theory, controlled by the primary producers themselves. In many cases producers have resisted the shift in the bal-

ance of power in the food economy to the manufacturers, either by gaining a stake in food processing themselves or by agitating for marketing boards and other forms of regulation to control private enterprise.

Another development in the agribusiness sector has shifted the balance of power within the entire food system: the tremendous growth in the control of the market held by the major retailers of food. This concentration of economic power has had an inevitable impact on the other "players" in the system: consumers, food manufacturers (large and small), and primary producers.

Emergence of the Modern Food-Retail Business: Chain Stores and Supermarkets

Bruce W. Marion (1986:274) argues that *two* major innovations have dominated the emergence of the modern retail food business: the chain store operation and the development of the supermarket.

The phenomenon of retail food-chain stores was pioneered by the Atlantic and Pacific Tea Company in the United States, which already owned two hundred stores by 1900 (Hampe and Wittenberg, 1964), and expanded rapidly until the Great Depression. The chain concept worked because by integrating into wholesaling, chains achieved greater buying power and efficiencies and were able to outcompete on price the small independent grocers with their much higher cost structures (Marion, 1986:294).

The Canadian pattern in the food-distribution business seems to have followed U.S. trends in the early period. The Report of the Royal Commission on Price Spreads of 1937 noted the rise of the retail food-chain store phenomenon in Canada, suggesting that it was imported from the United States, where it had originated in the tea trade around the middle of the 19th century. Although variety and drug companies were initially strong promoters of the chain concept, it was in the food sector that the largest expansion of chain operations took place in Canada during the boom years of 1925 to 1930. Indeed, of all of the chains operating in 1930, more than half had been established in the previous five years. The 1930 census estimated that of all the stores owned by chains in 1930, about 67 per cent were opened between 1926 and 1930 alone.[1] The greatest expansion took place in 1929, just before the Great Depression (Canada, 1933:4). By the mid-1930s the two largest food-chain companies in terms of sales – Dominion Stores and Loblaw – were also among the earliest chains to be established. The third largest chain – the Atlantic and Pacific Tea Co. – got started in Canada somewhat later. In the 1990s, some sixty years later, two of these three chains – Loblaw and A&P – are still in a commanding position in the food-retail business.

The tumultuous conditions of the Depression years had a strong impact on the food-distribution business in Canada. The 1941 census found that less than 3 per cent of the stores existing that year had been around in 1919 or earlier. Moreover, 65 per cent of the stores established before 1930 had closed between 1930 and the early 1940s (Canada, 1944:8,9). The economies of scale offered by the larger supermarket format were apparently a powerful motivation for change during these difficult times.[2] The average sales per store increased by 91.6 per cent over the decade, with most of the increase coming at the end of the 1930s. Among the chain stores, outlets with $200,000 or more of annual sales accounted for only about 6 per cent of all food sales made by chain stores in 1934. By 1937 chain outlets with more than $200,000 of sales had about 15 per cent of all chain store sales, but by 1941 their sales had increased to 51 per cent of the total dollar volume of food-chain stores (Canada, 1944:4).

Although the growth of the chain concept of retailing had been especially vigorous just before 1930, chains still only controlled about one-quarter of all grocery sales in Canada in that year. The rest of the sales were accounted for by firms – most of them necessarily small businesses – with less than four stores under any one owner. In fact, over 30,000 of the 32,523 retail food establishments enumerated by the census in 1941 were classified as independent operations. By 1941 the chains had increased the proportion of food sales they controlled, but only slightly to 27 per cent of the total. Nevertheless, the fact that the chains had this much of the market, with only about 5 per cent of the stores, shows the rather special nature of the stores they did control, even at that time. Almost three-quarters of the chain stores had annual sales of over $100,000 in 1941. Among the independents, only 10 per cent of stores did that volume of business annually (Canada, 1944:Table 5).

In the early period, retail food chains were more differentiated than they are now. Some chains specialized in "general line grocery" items, principally canned and bottled goods, along with other major items consisting of basic semiprocessed or unprocessed agricultural commodities: butter, cheese, sugar, fresh fruits and vegetables, and eggs. These few items made up about 85 per cent of the value of total sales. Notably absent were meats, which were still sold primarily through speciality meat stores. By the 1930s the chain concept had taken hold in the meat sector as well. Retail meat chains sold some general line grocery items, occasionally butter, cheese, eggs, milk and cream, and seafood, but 85 per cent of their sales came solely from meat and poultry.

The census also includes a third category of food-retail chain, this being the "combination" store chain. The main innovation of this type of operation was the inclusion of meats with the items offered by the typical grocery chain store. Even by 1930 these "combination" stores were the dominant type of chain operation, accounting for 58 per cent of the over $128 million in annual sales by all types of chains in Canada. "Grocery" chains had 35 per cent of total sales, while meat chains were relatively less significant, with about 7 per cent of all chain sales. What is remarkable, however, is the fact that even by this early date the stores belonging to four chains among all the "combination" chain operations accounted for fully 83 per cent of the total number of stores within this type of chain, and over 80 per cent of all sales.[3] In fact these four chain operations had very nearly one-half (48 per cent) of all the grocery retail sales in Canada accounted for by chain-type operations in 1930 (Canada, 1931:Table 7). Clearly, corporate concentration in the food-chain retail sector is nothing especially new in the Canadian context. The difference in the 1930s, of course, was that chain store corporations then had a much smaller proportion of all grocery sales in the economy than is the case today.

The major second innovation, the supermarket, began its existence independent of the chain stores and combined cash and carry, self-service, and a broad selection of products with an emphasis on low prices and high turnover. Home refrigeration and the extension of automobile transportation to an ever broader proportion of the public created the environment for supermarkets to flourish. It was not long before chain store companies adopted the supermarket innovation as well (Marion, 1986), further enhancing their position in the retail trade. By 1948 chain store operations had captured 34 per cent of grocery sales in the United States, with the largest chains, those with over fifty stores each, having 27 per cent of the market.[4] Almost 60 per cent of sales were still controlled by independent operators owning a single store, but by 1982, barely thirty years later, this situation had completely turned around. By then chains controlled about 62 per cent of all grocery sales (ibid., Table B-2), and wholly independent retail grocers had only 26 per cent of this market.

Food Retailers Today: In the Driver's Seat

The degree to which power in the Canadian food system has shifted to the retailer food chains may not be widely appreciated by those outside the food industry, including those in the media and academia, but it is no secret to most agricultural producer organizations and food manufacturers. It is difficult to say when this shift occurred, for the truth is that it has probably been taking shape for a number of years. Students of the U.S. and British food industries have been analysing the trend for more than a decade (see Cotterill and

Figure 6
Determinants of the Relative Bargaining Power of
Buyers versus Sellers

A buyer group is powerful when	A supplier group is powerful when
• it is concentrated or purchases large volumes relative to seller sales.	• it is dominated by a few companies and more concentrated than the buying industry.
• products purchased are standard or undifferentiated (i.e. commodities).	• it sells products with few substitute products for sale to buyers (highly differentiated products).
• buyers incur few costs when switching to alternative suppliers.	
• buyers pose a credible threat of backward integration.	• buyer is not an important customer of the supplying industry.
• supplier's product is unimportant to quality of buyer's products or services.	• supplier's products are an important input to the buyer's business.
• buyer has more complete information.	• supplier group poses a credible threat of forward integration.
• buyer can influence consumers' purchasing decisions.	

Source: Adapted from Michael Porter, *Competitive Strategy: Techniques for Analyzing Industries and Competitors* (1980), as reported in Hamm (1982:2).

Mueller, 1979; Hamm, 1981, 1982; Howe, 1983; Marion, 1984), drawing attention to the growing power of retailers in the food system in those countries. In Canada much less attention has been paid to the phenomenon, despite evidence suggesting that the trend has proceeded much further here than in the United States, at least.

Our understanding of the determinants of the relative bargaining power of buyers versus sellers of commodities in any given industry has benefited from landmark studies such as that by Michael Porter (1980). Porter argues that there are a number of key determinants of the relative economic power of *buyers* and *sellers* within our economy. These determinants are enumerated in Figure 6. We shall consider the structure of the Canadian food system in light of them.

Retailer Concentration: The Most Important Determinant

The most salient feature distinguishing the Canadian food-distribution business has to be the degree to which Canadians have allowed the wholesaling and retailing of food to be captured by so few corporations. The degree of concentration in this sector of the food economy is extraordinary. By 1987, for instance, the largest five grocery distributors in Canada accounted for about 70 per cent of all sales. In the United States, by comparison, the top five firms had only 24 per cent of total sales. In other words, the control of the top five firms in Canada was *two and a half times* that of the top five firms in United States. This level of concentration has increased relatively over time as well (Ferguson, 1992:1).

In recent years the more stringent U.S. anti-trust legislation has established measures to determine "unacceptable" levels of corporate concentration. In its 1984 guidelines for judging whether corporate mergers contravene the public interest, the U.S. Department of Justice adopted the Herfindahl-Hirschmann Index (HHI) to replace older measures of corporate concentration. With this index the spectrum of market concentration can be divided into three broad ranges: if the HHI is below 100 the industry is considered to be "unconcentrated"; if it is between 1,000 and 1,800 the situation is considered "moderately concentrated"; and if it exceeds 1,800 the industry is judged to be "highly concentrated." According to the U.S. Department of Justice's guidelines, if a proposed merger in an industry in the 1,000 to 1,800 range increases the industry's HHI by more than 100, or if a merger would increase the ratio of the top four firms beyond the 1,800 threshold, the department could challenge the merger (see Ferguson, 1992:6).

If we compare the HHI of the top five firms in Canada and the United States, it is fairly clear how much the corporate structure of our food-distribution business departs from the U.S. situation. In fact, the index of concentration in Canada is *eight* times greater than it is in the United States, and more than twice the 1,800 threshold that would trigger a challenge by government in the United States.[5]

Given that the largest five grocery chain firms in Canada accounted for some 70 per cent of sales in 1987, this is a remarkable degree of concentration in itself. But the national level is not necessarily the most useful level for measuring concentration, because the market within which competition takes place exists at a much more *local* level – basically the level of individual towns and cities. As one leading authority on U.S. food distribution notes, "There is widespread agreement that food retailers sell in geographic markets that are inherently local. Consumers do not travel from one region to another, or even one city to another, to shop for groceries" (Marion, 1986:302).[6]

If we wish to examine the level of choice that consumers realistically have in the market for grocery products, we need to look at the number of grocery chains operating in a particular town or city. At this level we find, for example, that the top three firms own or sponsor 82 per cent of all grocery stores in the city of Vancouver. In Calgary a single firm owns or sponsors 57 per cent of all stores, and in Edmonton one firm has 75 per cent, while the top two firms in Calgary and Edmonton control 81 per cent and 94 per cent respectively of all stores.[7] In Eastern Canada the situation is similar; for instance, three firms control 80 per cent and 91 per cent of all stores in Montreal and Quebec City respectively (Ferguson, 1992:12, 15, 21). At the city level, the level on which chains actually compete and formulate their strategic plans (Marion, 1986:302), the degree of concentration in the food-distribution business is stark indeed for a number of major metropolitan markets in Canada.

Concentration and Prices

With the industrial transformation of food, the domination of just a few companies in the market inevitably has its consequences – for primary producers, for other processors, for consumers of their products, and even for the rural communities dependent upon agriculturally related activities. How has corporate concentration at the food-retail level and the economic power this generally confers been manifested in the food economy?

First of all there is the relationship between retail concentration, the price we pay for food, and the matter of what proportion of the final price is appropriated by each sector of the food chain. In the United States several studies have found a significant positive relationship between retail food prices and supermarket concentration (Marion et al., 1979; Lamm, 1981; Hall, Schmitz and Cothern, 1979; Cotterill, 1984). In other words, the higher the concentration, the higher food prices tend to be. Does a high degree of distributor concentration help produce higher food prices in Canada? It seems reasonable to conclude that it would, given evidence from the United States supporting such a relationship. Indeed, the periodical *Food Market Commentary*, surveying some ninety-two food items in both U.S. and Canadian border towns, determined that food on the Canadian side cost 10 per cent more than food across the border. This would be little surprise to most Canadian consumers, who have witnessed much discussion about higher Canadian food prices in the media. But why is this the case? What is perhaps most significant is how this situation has been interpreted, and here the media have played a key role.

The media have paid considerable attention to the purported role of the supply management system and producer marketing boards in causing the prices of food to be higher in this country. As trade between Canada and the United States has been liberalized, and barriers to U.S. competition lowered, large Canadian food companies have been vociferously criticizing the primary production sector for supposed inefficiencies, which the food companies say are responsible for high raw material costs and in turn make it difficult for these firms to compete with U.S. companies. Food-industry organizations and their allies in the business press and the universities have launched a wide-ranging attack on producer marketing boards, and supply management in particular, as the source of our supposed uncompetitiveness.[8] Led by a series of articles and numerous editorials in *The Globe and Mail* in the fall of 1990, and echoed in much of the Canadian business press, these powerful critics have characterized the supply management system as a "wretched system" that "does not work" and "damages the competitive advantage of an entire sector of the economy" (Janigan, 1990:88,96). Because of supply management, it is said, "The price of basic foods in Canada is too high" (ibid.; Matkin, 1990; and *The Globe and Mail*, editorial, Nov. 17, 1990), and our food industry is uncompetitive internationally (Corcoran, 1991).

A major study of price spreads for farm commodities during the 1980s challenges this conclusion, in several significant ways (see Ferguson, 1991). Not only does it support the conclusion that retail concentration has contributed to higher food prices, but it also gives an indication of how much of the price increase of various commodities is added on by the retailer. In doing so it provides a valuable insight into the relative power of retailers vis-à-vis other sectors of the food system, and specifically producers and processors.

Firstly, a study of prices at the farm gate level and the processing (wholesale) and retail levels indicates that for several key foods, the proportion of price increase during the 1980s accounted for by the food retailer is substantially *greater* than the proportion accounted for by either the processor or the farm operator. In the case of beef, the net price increase between 1983 and 1990 going to the retailer was 26.6 per cent, while the increase going to the processor was 13.4 per cent and that going to the farmer, 8.8 per cent. In the case of cheese, the price increase going to the retailer was almost double the increase going to either the farmer or the processor. In the case of chicken, farmers received a price increase that was only about one-third of the price increase going to food-retail chains during the 1980s. In the case of turkey, farmers received less than half the increase going to retailers. The relatively low proportion of the price increase going to farmers producing cheese, chicken, and turkey is significant because these commodities are regulated by a supply management system, which has been held responsible for our relatively high food prices. Clearly, other factors are at play.

It might be argued that food retailers were justifiably receiving a relatively greater share of our food dollar because their costs have increased at a greater rate than those of processors, or farmers. However, Agriculture Canada's *Food Marketing Cost Index* shows that on this score the cost of retailing during the 1980s actually increased less than did costs for the processing of food (Ferguson, 1991:42). Moreover, if costs were a major factor in determining the increase in food retailers' prices, we would expect that higher costs would have prevented higher retail prices from being realized as higher retailer profits. Again, evidence suggests that this was not the case.

The food-retail business was the most profitable of all Canadian retail operations in the period 1986-90, including clothing retail, department stores, and specialty stores.[9] Evidence indicates that the profitability of the food-distribution business has been outstanding relative to many other sectors of the Canadian economy, and in particular relative to agricultural producers. Using Statistics Canada data on the return on investment (capital) for each sector (see Figure 7), it is noteworthy that the average five-year return of investment for Canada's major food distributors was 17.2 per cent, while for farmers it was 6.2 per cent; and the calculation for primary producers does not place a value on unpaid family and farm-operator labour. Had this been figured into the equation, the costs for producers would have been substantially higher, and their returns on investment correspondingly lower.

If the food-distribution business was actually characterized by genuine competition, and if producer marketing boards were as powerful in the marketplace as some critics suggest, then we would expect this state of affairs to have a negative impact on retailer profits. But the data indicates that retailer profits in the last half of the 1980s were in fact quite robust, to say the least. Moreover, a comparison of the returns of the subsidiaries of the two major U.S. food retailers operating in Canada – Safeway and A&P – with the returns of their U.S. operations sheds further light on the theory of the "inefficiency" of the Canadian agricultural production system as the main factor in the difficult time that processors are having competing today.

A&P's operations in Canada made 20 per cent more on each $100 of sales than its U.S. operations did. In the case of Canada Safeway, the comparison is considerably more dramatic. In Canada this firm had a return of $1.96 on each $100 of sales, while its U.S. operations had a return of thirty cents on each $100 of sales (Ferguson, 1992:50) In other words, Safeway's Canadian operations made 553 per cent more on each $100 of sales. We could ask, to what degree is this impressive showing in Canada related to the firm's extraordinary

Figure 7

Agricultural Producers' Return on Assets (Capital)
5-year average 6.2%

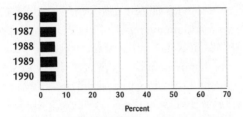

Source: Ferguson, 1992:30 (calculated from StatsCan).

Food Manufacturing's 1986–90 Return on Capital
5-year average return 16.91%

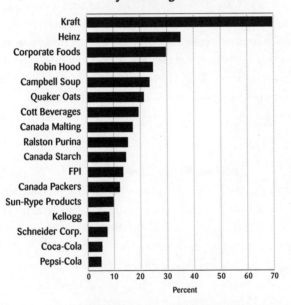

Source: Ferguson, 1992:30 (calculated from *The Globe and Mail Report on Business* Top 1000).

Food Distributors' 1986–90 Return on Capital
5-year average return 17.21%

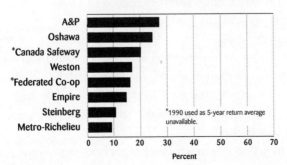

*1990 used as 5-year return average unavailable.

Source: Ferguson, 1992:31.

degree of control over several major metropolitan markets in Western Canada?

Finally, if we take all major food-distribution firms in Canada and the United States, that is, those with sales in excess of $1 billion, Canadian food-distributors in 1989-90 made on average 7.4 per cent more profit than U.S. distributors did (ibid., 48). It is indicative of the favoured position of the Canadian food distributors that even in the depths of the most severe economic downturn since the Great Depression, food-retail profits were still reasonably healthy, whereas corporate profits for the economy as a whole were decidedly depressed.[10]

This data suggests that real attempts to uncover the determinants of relatively higher food prices in Canada will have to look beyond a weakened and crisis-ridden farm sector for its answers. Perhaps the more interesting question is: why have the giant food retailers attracted virtually no attention from the major media on this question? Instead, the media have chosen to focus their attention elsewhere in the food chain, that is, on the institutional arrangements protecting primary producers – the sector that is least organized, most fragmented, and that has never truly emerged from the crisis it went into in the early 1980s.

Backward Integration

There are other determinants of retailers' power besides their economic concentration. Porter's model suggests that backward corporate integration into food processing by retail buyers, or even the credible threat of doing so, enhances their power over their suppliers. Marion's study of the U.S. distribution system shows that this strategy has been used by food retailers there for several decades. Retailer integration into manufacturing tends to be in major product lines such as fluid milk, fresh and processed meats, and bread products, which together accounted for two-thirds of the value of the food that the fifty largest retailers manufactured through their own facilities in the late 1970s. The proportion of their total food sales they manufactured themselves declined somewhat after the mid-1950s, from about 10 per cent to 7.6 per cent in 1977 (Marion, 1986:336). Marion notes that it has primarily been the very largest food retailers that are the most vertically integrated; that is, they produce the largest percentage of their total food-store sales. The largest four firms had increased the percentage supplied by their own manufacturing firms to over 10 per cent by 1977 (ibid., 337).

Backward integration into manufacturing has not been as prominent a part of the Canadian food-retail scene, with one major exception and a few partial exceptions. Canada Safeway is a partial exception, because in addition to being vertically integrated in the United States, it had its own food-processing concerns in Canada. The Canadian processors have been sold off. The major anomaly in the Canadian food system is Loblaw, which actually does not represent a case of backward integration, but of forward integration. In this case, in 1947 a nationwide firm primarily involved in the baking industry, George Weston Ltd., purchased the retail food-chain stores owned by Loblaw Groceterias and through several subsequent mergers with other chain operations built the largest nationwide retail food operation in the country.[11] The culmination of this long exercise in the accumulation of food-distribution operations is an extensive chain of 1,595 food-retail outlets (329 corporate and 1,266 franchise) integrated into regional wholesaling organizations also controlled by the parent firm, which has a substantial stake in the U.S. food-distribution business as well. In 1990 these chain operations brought the firm $8.4 billion in sales, a hefty portion of the total retail food sales in the Canadian market approximating $40 billion (*Canadian Grocer*, August 1991).

An examination of the corporate structure of George Weston Ltd. reveals why it is such a significant component of the Canadian food system (see Figure 8). The parent company, George Weston, in addition to extensive baking industry holdings, was by the 1960s expanding rapidly into many other sectors of the Canadian food industry, including fish processing (B.C. Packers, Connors Bros.); milling (McCarthy Milling Co., Soo Line Mills); dairy processing (William Neilson Co., Donlands Dairy); fruit processing (Bowes Co.); sugar refining (Westcane); and confectionary (Cadbury Schweppes, Willard's Chocolates).

Weston had also established itself as an international food corporation of considerable magnitude, with its principal expansion being in the United States and the United Kingdom. The U.S. expansion began in the 1940s with purchases of baking firms. Several decades later the company purchased such major U.S. assets as Stroehmann Bakeries and extensive food wholesale and retail operations, including Peter J. Schmitt Co. and National Tea Co. By the late 1980s, fully 33 per cent of the $11 billion in annual sales accounted for by all Weston Ltd. companies came from the United States.

While the Weston empire was expanding its hold on the Canadian food industry, it was also metamorphosing from an increasingly complex vertically and horizontally integrated food corporation into a *conglomerate* firm with holdings beyond the food sector *per se*. Here, a key event was the purchase of the

The Weston Family

George Weston
1865–1924
Baker and businessman

Willard Garfield Weston
1898–1978
Food merchant and manufacturer

George Weston

The current Weston empire, including Loblaw Companies, George Weston Ltd., and Holt Renfrew, among others, is currently run by the third generation of Westons. The business began in 1882 with the purchase of two bread routes in Toronto. In the years that followed George Weston continued to buy up more routes, and in 1897 he opened his first bakery, Model Bakery, which could produce up to 3,200 loaves of bread a day.

Three years after the opening of his Model Bakery, George Weston went into partnership with four other bakeries to form Canada Bread Co. At the same time he owned a biscuit factory but held it separately. The partnership lasted until 1920 when it was dissolved and Weston independently re-established his own bakery. In addition to his bakery interests, Weston was also interested in municipal politics. He ran for alderman in Ward Four in Toronto and was elected four times, but never became a strong force at City Hall.

George Weston died in 1924 of pneumonia, but several years before his death the bakery business saw a new force – Garfield, one of George's four children. The young Weston returned home from the First World War in 1919 and began working for his father. He worked in the biscuit factory and quickly moved up through the ranks. In two years he became vice-president of George Weston Ltd. and by 1924 he had been elected president and general manager.

Following his father's death, Garfield Weston began expanding the Weston name internationally by buying up small bakeries and biscuit companies in Canada and other countries, including the United States and United Kingdom. In 1928 George Weston Ltd. became incorporated as a federal, public company.

Garfield moved his family to England in 1933. And as his father had done, he participated in local politics. Garfield became a British Member of Parliament but, again like George, never played an influential political role. He spent almost a decade in Britain and during this time continued to increase the Weston empire. British organized labour was always suspicious of him and viewed him more as a speculator than anyone interested in the baking and food industry.

After World War II Weston moved his family back to Canada and continued with his massive acquisitions at home. He purchased the Winnipeg-based food wholesaler Western Grocers Ltd., Loblaw Ltd., and the William Nielson Co. His next strategy was to target family-controlled food wholesalers because they supplied food to independently run stores. The 1940s and 1950s provided an ideal time to squeeze the independent stores and wholesalers out of business and replace them with large supermarket chains. It was for this purpose that Garfield Weston bought Loblaw Groceterias, in addition to National Grocers Co. in 1955 and Atlantic Wholesalers in 1959. In North America he took advantage of new social trends that would enlarge his corporate empire. As author Charles Davies puts it, "The baby boom, rising incomes and massive population shifts from cities to suburbs all fuelled the change in shopping habits, pushing the supermarkets' share of all U.S. food sales from 35% in 1950 to 70% in 1960."

Outside of North America Weston continued to expand as well. He increased his company's holdings in Australia, New Zealand, South Africa, and the United Kingdom, and he supplied 10 per cent of their total bread sales from his bakeries. His growing corporate empire became increasingly complex and secretive. In 1966 the federal government of Lester Pearson, after much public pressure, established a federal commission to investigate the issue of price fixing. In an important finding the commission argued in favour of increased corporate disclosure to the public, with a particular reference to the Weston group.

In the 1970s Garfield Weston began to work less and retreated more into his private life. He began to pass more of the work onto his son, Galen. He died in Toronto of a massive heart attack at the age of eighty. The next generation of Westons continued to increase the company's profitability. Galen Weston would prove to be yet another marketing success story, helping to turn a slumping Loblaw's into a grocery store giant and continuing to make the Westons one of the richest families in Canada.

Garfield Weston

Source: Davies (1987).

Figure 8
The Weston Company: Vertical Integration in the Food System

W. Galen Weston

Wittington Investments Ltd.
(Canada)

George Weston Ltd.
(57% Controlled)
(sales in 1992, $11.6 billion; assets in 1992, $4.0 billion)

Food Distribution Group

(Profits in 1990, $212m)

Weston Food Distribution Inc./Distribution Alimentaires Weston (100.0%)
- Loblaw Companies Ltd. (76.7%)
  - Ad X Communication Inc. (100.0%)
  - Fortion's Supermarkets Ltd. (100.0%)
  - Hasty Market Inc. (100.0%)
  - Loblaw International Merchants Inc. (100.0%)
  - Loblaw Inc. (100.0%)
    - 353155 B.C. Ltd. (100.0%)
    - 496294 Ontario Inc. (100.0%)
    - A. Dionne and Fils Limitée (100.0%)
    - Atlantic Foods Ltd. (100.0%)
    - Brim Products Co. Ltd. (100.0%)
    - Busy B Discount Foods Co. Ltd. (100.0%)
    - Can-Dania Importing Co. Ltd. (100.0%)
    - CCG Security Inc. (100.0%)
    - Central Canada Grocers Group (100.0%)
    - Chignecto Holdings Ltd. (100.0%)
    - Dionne Limited/Limitee (100.0%)
    - Double Z Uniform Rentals Inc. (100.0%)
    - Extra Foods Ltd. (100.0%)
    - Extra Fare Supermarkets Ltd. (100.0%)
    - Foodwide of Canada (1977) Ltd. (100.0%)
    - Foremost Foods Ltd. (100.0%)
    - Golden Mill Bakery Inc. (100.0%)
    - Grand Union Markets Ltd. (100.0%)
    - H.E. Fisher Ltd. (100.0%)
    - Kelly, Douglas and Co. Ltd. (100.0%)
    - Kingsway Frozen Foods Ltd. (100.0%)
    - Loblaw International Holdings Inc. (100.0%)
      - Glenmaple Overseas B.V. (100.0%)

- Loblaw Quebec Inc. (100.0%)
- Loblaw Supermarkets Ltd. (100.0%)
- National Grocers Co. Ltd. (100.0%)
  - 136832 Canada Inc. (100.0%)
  - Hasty Market Holdings Inc. (100.0%)
- Sunfresh Ltd. (100.0%)
  - Swiss Chalet J.V. Corp. (50.0%)
  - Westcare Inc. (100.0%)
- Western Commodities Inc. (100.0%)
- Western Commodities Ltd. (100.0%)
- Western Commodities U.S. Inc. (100.0%)
- Westfair Foods Ltd. (100.0%)
  - 141836 Canada Ltd. (100.0%)
  - 422984 Alberta Ltd. (100.0%)
  - Econo-Mart Ltd. (100.0%)
  - Extra Food & Drug Ltd. (100.0%)
  - Product Awareness 86 Inc. (100.0%)
  - Shop Easy Stores Ltd. (100.0%)
  - Shop-Rite Stores Ltd. (100.0%)
  - Super Value Stores Ltd. (100.0%)
  - Tom Boy Stores Ltd. (100.0%)
  - Western Grocers Ltd. (100.0%)
  - Westfair Drugs Ltd. (100.0%)
    - Westfair Drugs (B.C.) Ltd. (100.0%)
    - Westfair Drugs (Manitoba) Ltd.(100.0%)
  - Willet Foods Corp. (100.0%)
  - Zehrmart Inc. (100.0%)
  - Ziggy's Food Inc. (100.0%)
- McKenna's Drug Store Ltd. (100.0%)
- Premier Packaging Ltd. (50.0%)
- Red & White Foods Ltd. (100.0%)
- The Suntex Food Group Inc. (100.0%)
- Wadland Pharmacy Ltd. (100.0%)
- Glenmaple Food Distribution Inc. (100.0%)

Eddy Paper Company in 1962. However, it was the acquisition of retail and wholesale food companies that transformed George Weston into what is now first and foremost a food-distribution company. Of its four divisions, food distribution, food processing, fisheries, and forest products, the food-distribution segment accounted for 78 per cent of total sales and 52 per cent of total profits by 1987.

It is its position as part of this highly integrated conglomerate firm, including a substantial food-manufacturing component, that gives the retailer Loblaw Co. such potent market power, not only vis-à-vis food manufacturers but also in relation to most of the other food retailers operating in the Canadian market.

Real Estate

Wittington Properties Ltd. (100.0%)
• Wittington Developments Ltd. (100.0%)
• Garrison Creek Ltd. (100.0%)
• Georgia Place Ltd. (10.0%)
• Wittington Properties (Montreal) Ltd. (10.0%)
  • Inglewood Realities Ltd. (100.0%)
• IPCF Properties Inc. (100.0%)
  • Westfair Developments Ltd. (100.0%)
  • Westfair Properties Ltd. (100.0%)
    • Westfair Properties Pacific Ltd. (100.0%)

Food Processing Group

(Profits in 1990, $68m)
• Bowes Co. Ltd. (Toronto) (100.0%)
• Brian Avon Sales (1986) Ltd. (100.0%)
• Cadbury Canada Marketing Inc. (100.0%)
• Canada Yeast Company Ltd. (100.0%)
• Chocolate Products Co. Ltd. (100.0%)
• Clark Diary (Quebec) Ltd. (100.0%)
• Dietrich Bakeries Ltd. (100.0%)
• Diversified Research Labs Ltd. (100.0%)
• Hampstead Food Industries Co. Ltd. (100.0%)
• Jade Foods Inc. (100.0%)
• Kennedy Agencies Ltd. (NB) (100.0%)
• McNair Products Co. Ltd. (Toronto) (100.0%)
• Nassau Diary Products Co. Ltd. (100.0%)
• Rose and LaFlamme Ltd. (100.0%)
• Saxonia Fruit Preserving Co. Ltd. (100.0%)
• Timmins Northland Bakery Co. Ltd. (100.0%)
• Wasco Foods Inc. (100.0%)
• Watt & Scott Inc. (100.0%)
• Western Pre-bake Ltd. (100.0%)
• Weston Bakeries Ltd./Boulangeries Weston Limitée
  (100.0%)
  • 2 for 1 Store Inc. (100.0%)
  • Boulangeries Weston Quebec Inc. (100.0%)
  • Cecutti's Bakeries (1983) Ltd. (100.0%)
  • Ready Bake Foods Inc. (100.0%)
• Veva Holdings Inc. (100.0%)
• William Neilson Ltd. (100.0%)

Resources and Other

The W. Garfield Weston Foundation
• Key Securities Ltd. (100.0%)
  • J.R. Booth Ltd. (99.1%)
• Wittington Investments Ltd. (23.8%)
  • George Weston Ltd. (54.2%)
    • 165841 Canada Inc. (100.0%)
    • La Baguetterie Inc. (100.0%)
    • Becco Investments Inc. (100.0%)
    • Weston Resources Ltd. (100.0%)
      • British Columbia Packers Ltd. (100.0%)
      • Canadian Packing Co. Ltd. (100.0%)
      • H.G. Helgerson (1964) Ltd. (99.8%)
      • Jager Fishing Co. Ltd. (100.0%)
      • McCallum Sales Ltd. (100.0%)
      • Nanceda Fishing Co. Ltd. (100.0%)
      • Nelson Bros Fisheries Ltd. (100.0%)
      • North American Testing Ltd. (100.0%)
    • Connors Bros Ltd. (100.0%)
    • Eddy Paper Co. Ltd. (100.0%)
      • E.B. Eddy Forest Products Ltd./Products
        Forestiers E.B. Eddy (100.0%)
        • Chaudiere Water Power Inc. (38.3%)
        • The Coulonge and Crow River Boom Co. Ltd.
          (38.4%)
        • Ormiston Mining and Smelting Co. Ltd.
          (15.0%)
      • Pineland Timber Co. Ltd. (50.0%)
    • Robichaud & Company Ltd. (100.0%)
    • Weston Energy Resources Ltd. (100.0%)
    • Weston Stadium Participation Ltd. (100.0%)
  • Wittington Financial Services Ltd.(96.7%)
  • Holt Renfrew & Co. Ltd. (100.0%)
  • Holt Renfrew Alberta Ltd. (100.0%)

Source: Statistics Canada, *Intercorporate Ownership,* 1992.

Developing the "Buyer's Label"

As Porter argues, retailers' power relative to their suppliers is enhanced to the extent that they can influence consumers' purchasing decisions. In the retail food business certain developments have not only influenced the control retailers have over consumer purchases, but also in general increased retailers' influence in the food system in other ways. This involves the phenomenon of the development of *private* and *generic label* products (sometimes referred to as the buyer's or corporate label), which are controlled by the retailers even though they often do not manufacture them.

As private label and generic products account for more and more food sales, retailer influence increases. In part, this is because these products squeeze the markets of the traditional brand-name products (Burck, 1979:71).[12] But it is also because these new products are typically produced by the medium and small manufacturing firms (Bitton, 1985:77; Connor et al., 1985:220-23). Retailers therefore usually have several suppliers to choose from, which is, of course, not the case with high-profile brand-name products (there is only one manufacturer of Maxwell House coffee, for instance). As one study notes regarding the manufacturing of private label and generic products, "Retail and wholesale buyers clearly sit in the driver's seat. Their power is probably greatest on products such as processed fruits and vegetables, which are seasonally processed and then carried in inventory for the rest of the year" (Connor et al., 1985:105). The enhanced retailer power with the advent of these new market strategies is thus passed down the food chain to processing firms, which come to experience additional pressures to "source" lower priced raw materials (Burck, 1979:72). Although few studies exist to demonstrate the point, it is unlikely that the farm sector has been unaffected by these pressures.

One Canadian firm in particular, Loblaw, has received widespread attention for the aggressive way it has promoted its buyer's label program in the last decade or so. Its innovation comes with promoting its own corporate label ("President's Choice") to the point where by the 1990s it had achieved the status of a powerful brand name in a wide variety of product categories. This move has great significance. Using an aggressive advertising program, Loblaw has challenged the brand-name products of all but the very largest food manufacturers in one product category after another. As a result, by 1986 it had purged some 2,500 brand-name products from its shelves (Fiber, 1986:B1). By the mid-1980s, 20 per cent of all the merchandise Loblaw sold was under its private labels. Its executive vice-president noted, "We're going to be building far above our 20 per cent level over the next five years – category by category. We should reach 30 per cent by then and over 50 per cent in the long term" (ibid.). He predicted that more and more of the manufacturers' brands would be displaced until only two main brands (Loblaw's being one of them) would be left to fight it out in any single product category.

Even factoring in a degree of exaggeration here, Loblaw's success in developing its house labels says much about the contemporary strength of retail versus manufacturing capital in the present-day food economy. We are well on our way to a situation where all but the manufacturers with the most heavily advertised brand-name products – largely a few U.S. and European multinationals – will be relegated to the status of private-label suppliers, typically under precarious conditions with little autonomy and small profit margins.

The alternative to this for some manufacturers has been the institutional trade, and for a few others direct sales to food-service chains and customers abroad have been possible.[13]

By 1992 Loblaw's success in promoting its corporate label had pushed at least one other retail giant in Canada to emulate this strategy. A&P, recently enlarged after it had swallowed the Ontario stores owned by Steinberg's, launched its "Master Choice" line of products to counter the initiatives of Loblaw's "President's Choice" line. In essence, with this development, the largest retailers are beginning to challenge the major brand-name manufacturers on their own territory. To the extent that they are able to win over consumer loyalties on these new retailer's label products, it will be away from the largest brand-name manufacturers. Retailers will then be able to command the premium prices historically charged by the largest brand-name processors, but they will not have to pay the same substantial costs in rebates and allowances. This saving could, in part, be passed on to consumers, giving the retailers' products a further competitive edge.

The importance of aggressively developing buyer's or corporate label products for strengthening the retailer's position in the food system has been indicated by Loblaw's President Richard Currie. "Corporate brands," he notes, "when linked to our computerized merchandising systems ensure the preservation of retail margins, which would be greatly diminished if we engaged solely in national brand-name discounting practices."[14] By developing their own products, retailers such as Loblaw are helping to ensure that they will be relatively insulated from the pricing policies of the largest food manufacturers.[15]

Shelf-Space Landlords:
The Role of Special Deals, Discounts, Allowances, and Rebates

One factor that tells us something about the relative power of retailers, a factor the Porter model does not consider, is the ability of buyers to command and receive special compensation from suppliers for stocking the manufacturers' products on their shelves. The size of this "compensation" or rent that retailers are able to extract tells us something important about the clout of retailers in our food system.

As D. Howe notes about the British food system, "The dominance of the large food distributors vis-à-vis manufacturers is evidenced by the additional discounts which these mass distributors are able to extract from processors despite the market concentration among the latter" (1983:113). Howe cites the case of the frozen foods industry, in which, despite a high degree of processor concentration (one firm has a 50 per cent market share), discounts to the largest retailers were reported to be "unavoidable" and to have exceeded "the cost savings in supplying them as compared with supplying the generality of customers" (ibid.).

In the United States retailers appear to have benefited very handsomely from the so-called "promotional allowances," to the point where some chains reportedly make up to one-third of their profits solely from trade payments meant for marketing (Shapiro, 1992:5). What of the Canadian situation, then, where food-retailer concentration is much higher, and so presumably is the ability to extract revenues of this kind from manufacturers?

In 1987 a food-industry analyst writing in *The Globe and Mail* brought to the attention of the public for the first time in many years the significance of this phenomenon in our food system. He estimated that the industry's practice of providing retailers with special discounts, allowances, and rebates was adding 10 to 15 per cent to grocery costs in Canada – though estimates of these expenditures have to be treated with some caution, because their hidden nature makes accurate measures difficult at best. The *Globe and Mail* report put the cost of these practices in the Canadian food industry at over $2 billion of the $32 billion of annual food sales at that time (Matas, 1987:1).

What is remarkable about these practices is that far from being a relatively recent development, they have been in place in one form or another for a very long time; and despite the negative implications they hold for other segments of the food system, and ultimately for the consuming public, very little has been done to curtail them. The 1937 Report of the Royal Commission on Price Spreads carried a detailed discussion of many of these practices and their damaging effects. The report was particularly concerned with the advantages that chain stores had over independent retailers with their power to demand special concessions from manufacturers and thereby provide themselves with a potent competitive advantage:

> But the development of large scale retail organizations in recent years and the disintegration of the jobbing trade in many lines have tended to make the matter of price discrimination more a question of volume of purchases than of trade status, of bargaining power rather than of possible savings in manufacturers' selling costs.
>
> A report made by our investigators covering 48 manufacturers supplying chain stores showed that in practically every case *the chain stores had a considerable purchasing advantage over independents through quantity*

and trade discounts, free goods, advertising allowances, etc. (Canada, 1937:225, my emphasis)

Almost fifty years later a provincial royal commission inquiry into discounts and allowances in the food industry of Ontario found that most of the practices mentioned in the Royal Commission report of 1937 were still in place, together with a number of new practices.[16]

Food retailers might be more accurately viewed as landlords renting shelf space to fewer and fewer food manufacturers. As one report argues, "The retailers charge the manufacturers to bring a new product into their stores, charge again for location, and space on their shelves and then charge once more for a prominent display in their advertisements and flyers. With almost insatiable appetites, many retailers then ask for additional payments over and above those already provided" (Matas, 1981:1).

It is testimony to the new-found influence of the retail chains that even the most powerful multinational food manufacturers have to play the rebate game. It is not a game without incidental costs, for as a *Harvard Business Review* study argues, such payments substantially increase the administrative costs of even the most influential processors (Buzzell, Quelch, and Salmon, 1990). As an example, its authors cite the figures given by one of the world's largest grocery product manufacturers, Procter & Gamble, to the effect that designing and implementing promotions to retailers absorbed 30 per cent of the management organization's time and 25 per cent of field people's time. However, as this study notes more generally about trade allowances, "Since there has been no noticeable decline of manufacturer and distribution profits, the consumer has presumably absorbed these costs" (ibid., 146).

The extraordinary promotional costs that are now central to the way food is merchandised in this country show no signs of going away. In fact, the opposite will more likely happen, because one of the latest developments in the distribution business is the invasion of the warehouse clubs. These merchandisers have been given a special boost by the rapid increase in special trade deals in the industry, because they are set up precisely to help to capture the limited range of brand-name products sold on special consignments, which are then resold in high volumes to consumers at prices that undercut the full-line resaler. Their growing presence in the industry has only made a bad problem grow worse (see ibid.).

Overall, the phenomenon of rebates, allowances, and special deals is one powerful indication of the major shift in the balance of power in the food system today. It is more than this, however. It is a powerful indicator that the food-distribution system as now organized contains profound irrationalities that in the end result in the squandering of the precious resources of the society at large.

The "Super" Retailers

Within the relatively small group of retailers that actually control the Canadian food-distribution system, a few are in a particularly favourable position because of the structure of their operations. In addition to the advantages we've already seen, these firms have certain characteristics that put them in a dominant position in the food-retail business as a whole. First of all, they are geographically diverse, spanning regions and countries. With this diversity they can weather regional downturns and benefit from regional economic upswings elsewhere in the continental economy. Second, these distributors typically encompass both the wholesaling and retailing dimensions of this business. Whereas in other sectors of merchandising, or more commonly in the food-distribution sector of the United States, retailers are confronted by large centralized wholesaling firms that can challenge the economic power of the retailers themselves, in the Canadian food-distribution business the largest retailers control their own nationwide wholesaling organization. In large part this is a strategy designed to reduce risk. As one major retailer notes, they have adopted the strategy "to operate in both the wholesale and retail sectors of the food distribution industry to minimize exposure to shifts in the balance of economic power between these two major components of the industry."[17]

A third advantage the largest firms have is their control of a range of store formats and ownership arrangements. The variety of store formats, from suburban "superstores" to small inner-city traditional supermarkets, allows these firms to capture various segments of the food-consumption market. Coupled with extensive real-estate holdings that are typically part of their asset base, they are in a favourable position to be flexible in changing store formats to take advantage of changing local demographic characteristics. In addition, the firms have generally moved extensively to convert their smaller stores into franchise operations that are supplied from their centralized wholesaling operations. In recent years these smaller stores were becoming increasingly costly to run, but the shift to franchise operations permitted major cost savings. A major factor here was replacing full-time unionized labour with a non-unionized part-time labour force in the franchised operations. This move to franchising has been extensive for some firms such as Loblaw Co., which by the early 1990s controlled over 1,200 franchise stores as opposed to only slightly more than 300 company-owned "corporate" stores (*Canadian Grocer*, August 1991:26).

The Irrationalities of Our Food-Distribution System

If we were to believe the food industry and its supporters in government and academia, then we in Canada have the best and most efficient food-distribution system in the world. The obvious abundance and immense variety displayed in the brilliantly lit, artfully decorated, spacious facilities that our merchandisers of food have built in all major population centres certainly reinforce this image. However, as with most things, all is not what it seems.

From time to time we get glimpses of the inside picture of the food-distribution business, and there is growing evidence of an extraordinary degree of what can only be described as social waste – valuable resources that are either squandered with no redeeming social benefit whatsoever, or else filched from the society as a whole and concentrated in the hands of a few. Of the resources squandered we might note the fact that in recent years, to counter the growing oligopolistic power of food retailers, the manufacturers of brand-name products have been forced to pour more and more revenues into advertising to shore up their brand-name images, just in order to bargain more effectively with the retail giants. As Connor, Rogers, Marion, and Mueller note, "The power of large buyers has probably encouraged some manufacturers to pour greater resources into differentiating their products. Rather than countervailing the most powerful sellers, the growth of large buyers may have an additional incentive for sellers to increase their market power" (1985:106).[18] Manufacturers do this because they know the alternative is to have their brands displaced from the shelves altogether and be forced into the much riskier and less profitable business of making private label products for the retailer. While this behaviour is rational from the point of view of the individual firm, from the point of view of the wider society our scarce resources are being wasted because we have allowed retailers to become too powerful.

The inside picture of the food-distribution business presents other examples where the existing corporate structure only serves to distribute resources from the relatively many to the very few. Powerful retail chain store operations, for example, have been able to use the so-called promotional allowances of food manufacturers to massively stockpile products on special. In the United States this has clearly become the normal way of doing business, and retailers have built immense warehouses solely to stock excess inventories of these goods bought on special (Shapiro, 1992:5). This practice of "forward buying," that is, of buying many months' worth of goods on special and then selling most of it at regular prices, was estimated by the *Harvard Business Review* study to account for as much as 50 per cent of the total stock of goods in distribution (Buzzell, Quelch, and Salmon, 1990).

By massively buying at low prices during a concentrated time of the year, retailers are forcing manufacturers to organize their production in far less than optimal ways, whereby production must proceed on an overtime basis for a relatively brief period in order to fill major orders, and then plants are run at less than capacity for several months until supermarkets replenish their inventories. A recent study found that this practice had resulted in $60 billion to $80 billion worth of excess inventory in the total U.S. food-distribution system (cited in Shapiro, 1992:5). There is, of course, a cost to carrying this extra inventory. Another study estimated that this extra inventory had added between $1.6 billion to $2.9 billion to food-industry costs (Buzzell, Quelch, and Salmon, 1990). It would be important to know whether there are just a few of these system-wide practices, or whether these are merely the "tip of the iceberg."

Some questions can and should be asked even without further evidence of the inside picture of the food-distribution business. The reflective observer cannot but question the erection of veritable food "palaces" where no expense is spared to provide the environment that will maximize the visual appeal and general attractiveness of the food we must buy, while at the same time Canadians by the tens of thousands are being excluded from the mainstream distribution process altogether. This has resulted in the necessary creation of an "alternative" food-distribution system, essentially because of the phenomenal growth of poverty in Canada. The national dimensions of this other food-distribution system, the most prominent aspect of which is the food bank, have received little attention to date. In Toronto, where we do have some good information, the number of people who were dependent on this other food-distribution system had swollen to 124,000 by 1991, and included 98,000 people receiving emergency food hampers and 26,000 people being fed in emergency hostels, drop-in centres, and similar programs.[19] About 51,000 of these people were reported to be children aged nineteen and under. Most disturbingly, the number of people dependent on emergency food programs increased in Toronto by almost 400 per cent in the period 1986-91 (DiManno, 1991).

These swollen and persistent numbers of people now forced to rely on the emergency food-assistance programs of Canada's "alternative" food-distribution system are resulting in pressure to institutionalize these programs following the U.S. pattern. In that country, instead of a heavy reliance on charitable donations through semi-annual "food drives," the bulk of the food now comes from food manufacturers who donate surplus or damaged goods in return for lucrative tax write-offs. Centralized food banks then "sell" the food they receive in this way to local agencies. A further aspect of this institutionalization is the development of a massive food stamp program that results in the expenditure of $12.3 billion annually to provide 18.7 million of the poorest Americans with daily sustenance (Pigg, 1989). This is surely evidence of a food system gone badly askew. The debate in Canada now is whether we will accept the levels of poverty and disenfranchisement that we currently have and move to the U.S. pattern of institutionalizing an alternative food-distribution system to service this poverty, or whether we will renew our efforts to change the economic system that systematically creates this poverty. These are very different choices.

We are told that we pay less on a per capita basis than most other societies for the food we eat, and yet this food is still too expensive for growing numbers of our citizens. We have little say in how our food is marketed to us, and although few would deny that it is pleasant to shop in the almost palatial surroundings of our largest supermarkets, only fools would believe that these surroundings are not somehow reflected in the price we pay for our sustenance. If there was more popular and local control over our food-distribution system – for instance, through a wide network of co-operative food stores linked to their own wholesaling enterprise – it seems quite possible that a different compromise could be reached between expenditures on the merchandising environment and the overall price level of the food that is the *raison d'être* for the industry in the first place. By alienating this business to an extraordinary few "merchant princes," we have excluded the possibility of providing broader access to the mainstream food-distribution system.

CHAPTER EIGHT

RESTRUCTURING AND CRISIS IN THE CANADIAN FOOD SYSTEM

•••••••

Sandra Watson and Anthony Winson

Darwin lives in business just as he lives in the jungle, and restructuring is an essential part of the process of renewal.
Canadian Business Magazine, November 1990

They didn't really care about anybody's feelings. They didn't treat you like you were human. You were a number to be discarded and nobody had to worry about you once you were gone. They didn't know you and didn't have to feel sorry for you.
Maple Leaf Foods worker reacting to
the closure of her plant, 1991

Our food system, for all its success in supplying increasing quantities of an ever-greater variety of goods, has also been characterized by both substantial waste and growing structural inequalities. These characteristics underlie a system that is not as rationally organized as we have been led to believe.

Until the 1980s these irrationalities were largely masked by the long-term dynamism of the wider economy, by the high prices farmers received for such significant Canadian farm commodities as wheat, and by the slow but steady growth in the demand for processed foodstuffs. By the early 1980s serious declines in farm commodity prices, combined with ruinously high interest rates, pushed many primary producers to the brink of disaster. Then, as these conditions persisted year after year, farm operators in considerable numbers went over the edge. They became caught in the tightening vice of ever-increasing input costs (including the costs of servicing their debts) and the stagnant or declining prices for the commodities they sold. Between 1980 and 1989, while farm labour costs rose by 43 per cent and other farm inputs increased by from 10 per cent to 30 per cent, the price of wheat in a loaf of bread, for example, declined by 22 per cent (see Ferguson, 1991). More and more farmers had to give up their land. This was done with great reluctance by many, and the rising tide of family stress, violence, and the ultimate message of despair – suicide – did not leave any rural farm community across the breadth of Canada entirely untouched.

By the late 1980s Canada was experiencing more than just a farm crisis. A powerful interplay of factors was at work, and the outcome of the interaction was being felt well beyond the farm sector itself. Serious stresses elsewhere in the food system, and most notably in the processing sphere, were in turn part of the wider shake-up of the Canadian economy, part of a corporate restructuring process that had begun to reshape Canadian society – and exact a high price for working people across the country.

The Material Conditions

The social dislocations that now characterize our food system are tightly bound up with qualitative changes in material conditions, accelerated technological developments, and a marked shift in the relative power of the nation-state and private capital in one country after another.

In Canada the development of the liberal welfare state after the Second World War involved, among other things, the entitlement of citizens to a much broader range of socio-economic benefits than had ever before been the case. In addition to a variety of state initiatives that could be classified under the rubric of "social welfare," there were important benefits gained by workers organized by strong industrial and public sector unions. These benefits were in the form of what has come to be called the "social wage": pensions, supplementary health care, maternity and child-care provisions, and severance packages, for example. In return for wages that allowed workers to participate in the consumer society, and the safety net supplied by benefit packages, unions "delivered" to the corporations a relatively peaceful and stable workforce that took some of the risk out of doing business.

By the 1970s this "social contract" between labour and capital had begun to break down. Among the immediate reasons for this was the declining profitability of the leading sectors of big business. The net rate of return on capital in the countries of the European Economic Community (EEC), for example, dropped by more than one-half between 1960 and the early 1980s (Warnock, 1988:Figure 1). At the same time the technological base of our society was undergoing a major qualitative change with the rapid introduction of silicon-chip technology into the world of business. This technology allowed business to be conducted in a way that had never before been possible. It also offered corporate capital the possibility of dramatically reorganizing its operations to reduce costs, especially labour costs, and enhance sagging profits.

Typically these cost savings have required the spatial relocation of corporate operations, especially to low-wage areas within the domestic economy, or to offshore locations, usually in the Third World. However, the mobility of capital – much enhanced by micro-chip technology – was hampered by a myriad of structures erected by both national governments the world over and well-organized labour movements. The interests of corporate capital could not be well served as long as the obstacles to restructuring were as strong as they were. A dismantling of these obstacles required that they be attacked at all levels, that their legitimacy be put into question, and that a political vehicle be set in motion to facilitate the agenda of removing the barriers to a massive economic reorganization. In this endeavour the efforts of a variety of groups – intellectual "think tanks," formal business groups and parabusiness organizations, and high level government-business consultative bodies – have played their role. This collectivity of interest groups has become known as the "new right." The people in these groups, as Patricia Marchak notes in her book *The Integrated Circus*, grasped the historical situation and through tireless effort on a variety of fronts successfully propagated a rationale for a particular kind of change in our social order: "The new right was not a passing convention of amateurs, nor was it, despite the temporary popularity of supply-side theorists and libertarians, a utopian movement. It was a well-funded global political organization. Business leaders of the OECD countries had an agenda long before the agenda became known as the new right's; already, in the establishment of the Trilateral Commission and several think-tanks throughout the industrial countries, corporate capital was organizing forums for disseminating its message" (Marchak, 1991:9).

Another Canadian researcher, Philip Resnick, distinguishes three currents of the new right in the North America context. The first is associated with the Trilateral Commission and reflects the views of the largest industrial and financial corporations. A second is associated with leading "liberal" economists such as Friedrich Hayek and Milton Friedman who, in their call for the freeing up of market forces, advocate a drastic rollback in the functions of government. The pursuit of a balanced budget becomes an overriding obsession for this group. A third powerful force within the new right is the fundamental religious organizations, the self-described "moral majority." As Resnick notes, in general the new right "involves individualism against collectivism, and repudiates the principle of equality.... It rejects the redistributionist ethic of the welfare state and the interventionist role of government" (1989:106).

The link between the amorphous forces of the new right and the increasingly globally oriented corporate capital is not always direct. Their interests are not exactly the same. Milton Friedman of the University of Chicago has been one of the most articulate and skilful users of the mass media for promoting his brand of libertarian, anti-statist economics. Although Friedman's standard target has been government intervention and fetters on a supposedly "free" market, he has on occasion publicly denounced large corporations and their propensity to develop a dependence on state subsidies.[1] Nevertheless, it is no accident that new-right think-tanks receive substantial funding from the largest corporate organizations. Most of the time, the message of the think-tanks has served the economic interests of the corporations.

As Marchak points out, while words alone cannot change the world, those words and ideas that are assiduously propagated and backed by heavy funding have great power (1991:11). During the 1980s, governments strongly influenced by the thinking of the new right were established in a number of major capitalist countries, including Britain, the United States, and Canada. These governments have successfully implemented policies that have significantly altered how their societies are organized. Key changes in the economic landscape have opened up these economies to unfettered foreign investment, to the dismantling of much of the regulatory apparatus built up over past decades, and to the privatization of leading state-owned industries, especially in the United Kingdom and Canada where public-sector investment was more prominent than in the United States. In Canada the change in the approach to foreign investment, for instance, came with the electoral victory of the Progressive Conservative party in 1984 and its quick dismantling of the Foreign Investment Review Agency, which had vetted foreign takeovers in Canada, and its replacement with Investment Canada. This signified a shift from at least some effort to scrutinize the impact of foreign takeovers to a wholesale promotion of foreign investment, in whatever form it takes. On the extent of British privatization, Madsen Pirie, an English economist and close advisor to Margaret Thatcher's Conservative government, noted, "In eight years ... we have privatized the ports and the docks. We have privatized the tele-communications industry, the radionics industry, the petroleum industry and North

Sea oil extraction. We've privatized the state bus companies and the state ship-ping lines. We've privatized shipbuilding in Britain, Jaguar cars, Leyland buses, Leyland trucks, freight haulage, the telephone service, the state airline, the aero-space industry, the state gas service, to name but a few" (1989:110).

Another major and more recent factor of this agenda has been the push to remove the variety of national and interprovincial trade and tariff barriers that had provided a relatively protected national market for indigenous firms. At the social level this agenda has entailed prolonged attacks on what have been the basic entitlements of citizens, including everything from food-stamp pro-grams in the United States to the universality of pension and family allowance benefits, and health care, in Canada.

Crisis in the Food System

As one food-industry journal stated, "The amount of rationalization and con-centration in the Canadian food manufacturing and distribution business is at historically high levels." (*PDR Notes*, July 8, 1992). The process of corporate con-centration and centralization has, as we've seen, been with us for a long time, and in recent decades it has impacted negatively on a wide range of players in the system. In Ontario, for instance, the number of fruit and vegetable proces-sors peaked at around 240 just after the Second World War and declined stead-ily until the early 1980s when it dropped precipitously to below fifty firms (Winson, 1990:390). Dairy-processing plants have evaporated across the Ontario landscape in the last twenty years, with the loss of 500 plants. By 1991 only 101 remained (*PDR Notes*, July 8, 1992). The decline of the small and medium food processors is nothing new, and the same could be said for the smaller retail food-chain operations and certainly the independent retailers. However, the situation since the late 1980s represents a qualitative break from the past. In the late 1980s the constellation of technological revolution, specific macro-economic policies, and the already highly concentrated corporate structure of agribusiness began to spawn the current phenomenon of corporate reorganization – a phe-nomenon usually referred to by the euphemism of "restructuring."

It is difficult to disentangle this phenomenon of restructuring from the massive merger activity that characterized the late 1980s in Canada and the United States. The mergers have precipitated much corporate "rationaliza-tion" – if only to eliminate the considerable duplication that typifies merged corporate entities. The recent takeover by Imperial Oil of Texaco Canada clearly illustrates this reality. The two companies both had vast refinery and retail capacities, and the merger brought on a great shedding of the duplicate operations, so that Imperial's total workforce shrank from 15,000 to 10,000 employees after 1989. By the early 1990s it was becoming painfully apparent that much corporate restructuring is simply a form of fall-out from the excesses of the previous decades, when leading companies built up bloated corporate empires on borrowed money – enterprises that could not withstand the pressures of an economic downturn. The 1991-92 collapse of the Reichmann empire and its Olympia & York Developments is a stunning example of this, and in the food system the recent moves by the "troubled" Edper Bronfman group to divest itself of the food empire controlled by its Labatt division illustrate this phenomenon (see *The Globe and Mail*, Feb. 13, 1993).

But what does "restructuring" actually entail? A technical definition of this process would emphasize a firm or industry's need to reorganize or refocus its operations, through sales of assets, plant closures, downsizing and streamlining of operations, and perhaps even new investments in keeping with an altered corporate strategy. In all but a few exceptional cases, the process has been synonymous with plant closures and job loss. While the business press and mainstream periodicals have devoted considerable attention to the shake-up in the corporate world, and in particular to the ups and downs of members of the Canadian corporate elite such as the Reichmann brothers and Robert Campeau, very little attention has been paid to the profound sociological dimensions of this phenomenon for Canadian society today. In terms of the food system the changes have had massive repercussions for the social fabric of the rural community, and of the rural family, bringing dramatic dislocations in the economic base that sustains them. In the last analysis, we are speaking of how people cope with a sudden termination of the activity – productive labour – that has for a long time given a focus to their day-to-day lives.

With the onset of trade liberalization and the open-door policy to takeovers by foreign corporations, the shake-up of the Canadian food industry has accelerated at a dizzying pace. It has coincided with, and some would argue it has gravely aggravated, the long-expected downturn in the business cycle evident by 1990. The province of Ontario was particularly hard hit by the repercussions of the economic depression and the Free Trade Agreement. According to one source monitoring the food industry, there were eight food-industry plant closures in 1989, thirteen in 1990, and at least twenty-three by the third quarter of 1991. Along with these forty-four plant shut-downs went 5,552 industrial jobs.[2] These firms formed part of the industrial base of the communities they were located in. When this industrial base is disturbed or substantially eliminated, sooner or later economic tremors are felt through every facet of community life – because superimposed on this manufacturing base are support and service industries, ultimately dependent on the base for their viability and expansion.

A particular case study – the story of Maple Leaf Foods – illustrates the contemporary reality of corporate rationalization and plant shut-down in the food system. The case involves the entrance of a British multinational food corporation into the Canadian economy in the context of the removal of restrictions to foreign capital by the Conservative government. This firm entered the economy through takeovers of Canadian food firms, first and foremost Canada Packers, once the nation's leading food company. What followed was a process of radical corporate surgery and disruption that spread from corporate headquarters to small rural communities.

The Real World of Restructuring:
Foreign Takeover and the Making of Maple Leaf Foods

Hillsdown Holdings PLC, a British company based in London, dramatically changed the complexion of the meat and food-processing industry in this country after its successful takeover of two Canadian food giants, Canada Packers and Maple Leaf Mills. Hillsdown Holdings was formed in 1975 by Harry Solomon, a British lawyer, and his old legal client, David Thompson, a meat producer. The company did well and became public in 1985. In 1987 Thompson left the company, selling off most of his shares (*The Globe and Mail*, May 4, 1990).

By 1990 Hillsdown was a diversified food processor in the top ranks of European food companies. Its after-tax earnings in 1989 were £149 million (or $286 million) on revenue of £3.7 billion. Today Hillsdown is Europe's largest fresh-meat supplier and fifth-largest food company. It is a private company with interests not only in food processing and distribution, but also in office stationary and equipment, furniture manufacturing, and timber and distribution. Hillsdown produces poultry, eggs, animal feeds, and fresh meats and bacon (*PDR Notes*, Aug. 7, 1987).

In 1987 Solomon turned his interests to Canada. He purchased the Toronto-based Maple Leaf Mills, a diversified and profitable agricultural products company, from Canadian Pacific for $361 million. Maple Leaf consisted of operations in flour, grocery, rendering, poultry, grain, and feed in Canada, the Caribbean, and Northern Ireland. In total it had thirty plants, many bakeries, four grain terminals, and twenty-three country elevators (ibid.). Solomon also had other smaller interests in Canada. He acquired a minority holding in the

Canadian fish company, Clearwater Fine Foods Ltd., in 1986. A year later he used Clearwater Fine Foods to purchase Blades Group of Canada, the country's largest herring producer.

Hillsdown revamped Maple Leaf Foods' operations into three groups: consumer foods, milling and baking, and agribusiness. Consumer foods now included prepared meats, grocery products, food service, and prepared poultry. The milling and baking group included flour milling, retail franchising, and international operations and became the country's largest manufacturer and distributor of fresh bakery products. The last segment, the agribusiness group, included fresh poultry and pork, hatcheries, international food trading, rendering, seafood processing, and the Shur-Gain division (Maple Leaf Foods *Annual Report*, 1991:13-23).

Hillsdown saw Maple Leaf as a vehicle to launch products into the North American market. In an interview Solomon was quoted as saying, "We've kept our secondary fishing interests in Canada but we would like to use Maple Leaf Mills as our main way into the North American market" (*Financial Post*, March 15, 1990).

When Hillsdown Holdings PLC purchased Maple Leaf, its first goal was to rationalize. Solomon gave David Newton, a nine-year veteran with Hillsdown, the job of restructuring the company. Newton oversaw plant closures and sell-offs, downsizing head office staff and organizing plant lay-offs. He sold its grain division to Cargill and its feed division to Robin Hood Multifoods (*PDR Notes*, April 1989).

Maple Leaf Mills and Robin Hood decided to swap some of their resources. Maple Leaf Mills sold its ten Master Feed animal-feed mill facilities to Robin Hood and in turn purchased two poultry processing plants from Robin Hood. As a result of this trade Maple Leaf Mills became the second largest poultry processor in Canada. In 1990 the company spent $23 million on an integrated bakery in Ontario with the intention of shipping some of its products to the United States. It expanded its Port Colborne plants to meet the growing market needs for "fresh-baked" bread, muffins, and cookies (*PDR Notes*, February 1989); and it decided to close its west-Toronto plant.

With the purchase of Maple Leaf Mills and later Canada Packers, Hillsdown wanted to expand its potential market outside of the United Kingdom, intending especially to take advantage of the free-trade agreement between Canada and the United States: "Hillsdown is currently dependent on the British market, where it cultivates, processes and distributes a wide range of food products. It aims to earn a third of its revenues in North America and a third in continental Europe within three years" (*Financial Post*, March 15, 1990).

When Hillsdown purchased Maple Leaf Mills it continued to look for other investments in Canada. Canada Packers was the largest food manufacturer and meat producer in the country since its formation in the 1920s under the auspices of J.S. McLean. When Canada Packers came up for sale it was an ideal buy: a diversified company with operations in many areas – poultry, beef, edible oils, grocery products, rendering, aquaculture, and food service. It had processing and manufacturing facilities in Canada, the United States, and Mexico and trading facilities in the United Kingdom, Europe, and Far East. During its peak it employed up to 14,000 people (*Financial Post* Yellow Cards, October 1990). Canada Packers had a long-established reputation and carried a popular label – Maple Leaf Brand – and most importantly it had a good distribution network. It could be used as a stepping-stone into the U.S. market, and it had assets that could be liquidated to help pay off Hillsdown's debts (*Financial Post*, April 18, 1990).

Hillsdown's first step was to merge its two Canadian subsidiaries to form a new company, Maple Leaf Foods Inc. It wanted to create an international food conglomerate that would find a large market in North America, specifically the United States. Next, Hillsdown set out to reorganize the new company, to eliminate duplication, divest itself of businesses that were not considered profitable, and in general restructure with a different focus. David Newton, who had been given the job of streamlining Maple Leaf Mills, was appointed as president and CEO and given the new task of restructuring the old Canada Packers.

Hillsdown has argued that low profit margins, overcapacity, and losses in the industry hurt Canada Packers considerably over the years. It has attacked the Alberta government for providing financial support to competing packers, such as a $6 million grant to Lakeside Packers in Calgary, a loan to Centennial Packers in Calgary, and its ownership and subsidization of Gainers in Edmonton. It argues that the company cannot compete with government dollars: "The continued involvement of the Alberta government in actively supporting our competitors and putting taxpayers' dollars up against shareholders' funds provides further reason for our planned action" (Canada Packers *Annual Report*, 1991:5).

The Entrance of Cargill

The beef-slaughtering business in Canada has been seriously affected by the actions of another foreign multinational giant as well. Its entrance into the Canadian market has undoubtedly had an impact on the direction and pace of

Hillsdown's retreat from this sector and the overall dislocation that has been the result. The new element to the equation was Cargill Inc. of Minneapolis, the largest privately owned company in the United States and one of the largest in the world. Cargill is a global commodity trader and the second-largest meatpacker in the United States. Its Canadian subsidiary, Cargill Ltd., is primarily involved in grain trading and handles about 8 per cent of all Western grain (*Financial Post*, Jan. 23, 1991).

In 1987 Cargill Canada announced that it was planning to build a fully integrated beef-packing plant in Alberta. It decided on High River, south of Calgary, because 80 per cent of the fattened cattle produced in Canada were finished in Alberta, and most of these came from close to this area (Kneen, 1990:64). Two years later the company built a $55 million state-of-the-art operation. The new facility included a slaughter floor and processing and packing facilities that could process up to 12,000 head of cattle a week and provide 500 jobs (*Financial Post*, March 10/12, 1990). The provincial government in Alberta provided it with a $4 million grant (Kneen, 1990:64). This was the first time Cargill had ventured into the Canadian meat industry (in the United States Cargill's subsidiary, Excel, with IBP and ConAgra, produces 82 per cent of the boxed meat). This move helped to reverse the trend that began in the mid-1970s of plant closures in Alberta's meatpacking industry (*PDR Notes*, May 4, 1987). As a result, Alberta would become a leader in meatpacking in Canada, with Cargill as one of the major players.

Cargill's presence in Alberta has had a tremendous impact on the meatpacking industry. Even before the plant was built Cargill contributed to the lowering of wages in other unionized companies. Its set wages were lower than Canada Packers plants in Alberta, a factor Canada Packers used to force wage concessions in 1988 under the threat of closing completely.

In 1991 Maple Leaf Foods decided to close its beef-slaughtering plants in Alberta and Saskatchewan and leave the fresh-beef industry entirely. It argued that it was leaving the industry because of the Alberta government's decision to subsidize its competitors, one of whom was Cargill. As a public company it could not compete with government money, and it could not compete with Cargill – with either Cargill's new technology or its deep pockets. It seemed a good time to get out while it still had a choice. According to one analyst, "Using aggressive pricing tactics, Cargill used that plant to undercut the established players to such an extent that it was partly responsible for Maple Leaf Foods Inc. decision to exit the beef business and leave Alberta" (*The Globe and Mail*, Aug. 20, 1992).

Cargill's entry into the industry, as well as its policies, created problems: "The meat-packing industry in western Canada has been in a turmoil since Cargill Ltd. built a $55-million state-of-the-art slaughterhouse and packing plant in High River, Alta., in mid-1989. The Cargill plant has contributed to the extensive overcapacity in the meat-packing business in the western provinces and the company's aggressive pricing policies have undercut many of its competitors" (*The Globe and Mail*, Feb. 21, 1991). Cargill has been accused of predatory pricing – it offers high prices for cattle and then sells the product below the going rate, a practice that gives the company an abundance of cattle and increases its market share, thus undercutting its competitors. Cargill can afford to lose money in the short term while aggressively seeking market share because of its deep pockets: its first public figures on the firm's revenues revealed a profit of $450 million (U.S.) on sales of $49 billion worldwide (*The Gazette* [Montreal], Aug. 20, 1992). These practices make it difficult for smaller companies, which can be just as efficient, to compete in the industry.

The Continental Strategy

David Newton oversaw all the operational changes and spent two years as CEO and president, altering Maple Leaf Mills and Canada Packers considerably. Once this was complete he was replaced by Charles Bowen, a marketing specialist knowledgeable about the United States. After the restructuring was complete, the Maple Leaf Foods focus was increasingly oriented to expansion south of the border, especially in the areas of bakery, grocery products, and poultry. Maple Leaf joined with ConAgra Inc., one of America's largest agribusiness companies, to form two flour-milling joint ventures that shared technology and profits. Maple Leaf agreed to contribute production from three of its mills in Canada, and ConAgra agreed to contribute production from its mill in Buffalo, New York (Maple Leaf Foods *Annual Report*, 1991:4-5). Maple Leaf reasoned that its milling operations were unprofitable and that the joint venture would increase its profitability as well as give it access to U.S. efficiency. Duplication would be eliminated – and thus more jobs would be lost.

The impact of these strategic alliances was to enlarge the companies' market share in different industries. It also helped make them more efficient by reducing duplication and costs, and it gave them more opportunities to spring into the United States. In a very short time Hillsdown changed management and operating styles, refocused its mandate, rationalized to improve efficiency and cost, co-operated with other companies, and tried to break into the U.S.

market. All of these changes increased its profits, but at considerable cost: not only thousands of jobs and careers, but also the permanent loss of two large and powerful Canadian businesses to an international holding company.

The Impact on Plants and Labour

Hillsdown radically reduced the size of its head office from 220 to 25 people (*The Globe and Mail*, May 4, 1991), sold off and shut down businesses, laid off thousands of employees, sold off assets to pay off debt, put in a new board of directors, and drastically changed the management structure to become more decentralized. The rationale behind these steps was to increase profits through improved margins and higher volume sales, and to increase efficiency. Hillsdown changed the direction of Canada Packers from a food processor and fresh-beef manufacturer to that of an exclusive food processor, increasing its value added potential and its number of brand-name products.

The difference between the management of Canada Packers under the McLean family and the strategy of its new British masters highlights the changes that have inaugurated this new era of "predatory capitalism." Under its previous owner, Canada Packers was run more along the lines of a family business. Although there were some segments of the operation that were not highly profitable, the company had not been in the habit of rashly disposing of operations in order to maximize profit in the short term. Overall the company had never lost money in sixty-three years. Immediate short-term gains were less of a priority than consistent growth over the longer term. Without the same interest in the employees or the communities its plants once supported, Hillsdown rationalized the operations, which meant thousands of lay-offs, plant closures, and a drastic downsizing of middle management.

One of Hillsdown's major strategies was to cut out and eliminate entire areas of the company. In its two-year restructuring period it either sold off or closed down plants or entire divisions. In December 1990 it sold Squirrel Peanut Butter, and early the following year it sold its Black Diamond cheese division to John Labatt. It also sold off the edible oils division, which had an annual sale of $250 million, to a joint venture company of Central Soya of Canada Ltd. and CSP Foods Ltd. (*PDR Notes*, Feb. 7, 1991). It also sold off all its salmon farms and six of its poultry plants, and completely shut down its fresh-beef unit.

Canada Packers was drastically restructured and rationalized. Its mandate was redefined away from its beef-slaughtering and packing-house image to that of a more value added food-processing company. It got out of the beef-slaughtering business altogether, closing down and selling operations in Calgary, Lethbridge, and Red Deer, Alberta, and in Moose Jaw, Saskatchewan. The most significant impact of this merger has been on labour. In the first twenty months of operation, twenty plants were either closed or sold off, and several thousand jobs were lost across Canada (see Table 6).

The Human Dimensions of Corporate Restructuring

There has been much written about corporate restructuring from the point of view of the corporate players, and of the investment community in general. There has not been much written from the point of view of those least able to adjust to a shift in corporate priorities – plant workers, their families, and the communities sustained by their jobs. To give voice to the experiences of these people, and to their concerns about the future, we will look into the story of a plant owned by Hillsdown Holdings and located in a small rural community in Southern Ontario – a community that, like many others across the country, has a narrow industrial base and is extremely vulnerable to the effects of a shut-down.

A poultry plant, Clearvalley Farm, located outside a small village northwest of Toronto, provided employment for people in the surrounding area.[3] The factory produced value added processed-chicken products for sale to restaurants, food services, and retail markets across Canada (Canada Packers "Annual Information Form," March 1991:11). Clearvalley was built in 1962 by a local farmer and his partner, and in 1981 the business was sold to Maple Leaf Mills. It changed owners again when Maple Leaf Mills and Canada Packers were bought by Hillsdown Holdings PLC and merged.

Hillsdown Holdings closed down the plant on June 14, 1991. At the time there were about 130 people employed at Clearvalley Farm. Most of them were women with at least five years of seniority, and they were from the local community. Hillsdown gave the employees eight weeks' prior notice (as required under the Ministry of Labour's employment standards). The reasons cited for the shut-down, according to a news release, were the age and layout of the plant and the high cost required to upgrade a controversial waste-water disposal system. For the past ten years the plant had been dumping its waste water onto surrounding farmers' fields. Some of the local residents were concerned that wells and a gorge behind the plant were being contaminated. The Ministry of the Environment began to put pressure on Clearvalley to find an alternative way to dispose of its waste water. In the end, deciding it was not worth the investment, Hillsdown Holdings shut down Clearvalley Farm and moved the operations to its Brantford plant.

One of Clearvalley Farm's employees, Betty, was then a young woman in

Table 6
Maple Leaf Foods Inc. Canadian Plant Closures
1990–92

Division	Location	Date	Employees affected
Canada Packers	Toronto, Ont.	1990	51
Canada Packers	Manitoba	1990	85
Canada Packers		1990	130
Canada Packers	Harriston, Ont.	1991	109
Canada Packers	Mount Forest, Ont.	1991	29
Canada Packers	Bramalea, Ont.	1991	125
Maple Leaf Mills	Elora, Ont.	1991	130
Canada Packers	Brampton, Ont.	1991	100
Canada Packers	Calgary, Alta.	1991	260
Canada Packers	Lethbridge, Alta.	1991	132
Canada Packers	Red Deer, Ont.	1991	130
Canada Packers	Aurora, Ont.	1991	100
Canada Packers	Sussex, N.B.	1992	120
Hoffmans	Kitchener, Ont.	1992	285
Canada Packers	Chatham, Ont.	1992	60
Canada Packers	Brantford, Ont.	1992	145
Canada Packers	Toronto, Ont.	1992	40
Canada Packers	Montreal, Ont.	1992	425

Sources: Maple Leaf Foods Inc. news releases, 1991, 1992; N. Heisler, "Food Processing Closures in Ontario," 1991; United Food and Commercial Workers Union, "Hillsdown Holdings Adjustment Practices," mimeo.

her twenties, married with three young children. After leaving high school she took a job at Clearvalley and started working on one of the lines. She was a lead hand at the time of the shut-down. She said she enjoyed working in the plant and liked her co-workers but was not completely sorry about leaving factory work. Betty was fortunate enough to find reasonable work a few months after the shut-down.

Hanna was a young woman in her early thirties. She lived common law with her partner and had two school-aged children. At the time of the shut-down she had five years of seniority, working as a lead hand and line-service person. When her job was terminated she decided to go back to high school. When we interviewed her she was enrolled in a year-long college course and still needed to finish grade 12. She hoped to find employment as a machinist in her community after completing the course. She recognized that she was one of the few employees who had the opportunity to go back to school and "better herself," as she said. She had experienced some economic problems and found that her life was noticeably disrupted with the lay-off.

Connie, a young woman in her twenties with two small children, also lived common law with her partner. She worked at the plant for seven years and was a lead hand when laid off. She decided to upgrade and complete high school when she lost her job. Connie was lucky to have found a job one week before her unemployment insurance benefits ran out. She was working in a factory again, but expected to be laid off again in another month. She said she really enjoyed her job at Clearvalley and was upset about having to leave the plant and her friends. She was bitter and critical of the company and the reasons for the closure. She and her family had experienced the financial and emotional strain of her job loss, but she believed the worst was possibly yet to come if she lost her current job and could not line up another one.

Rita was a middle-aged woman in her mid-forties. She was married for the second time and had two adult children. She worked at Clearvalley Farm for twelve years and was employed by all three of the different owners. She started working at the plant following her divorce from her first husband, re-entering the workforce after staying home for several years. At that time she was a single parent with two children, but when we interviewed her she was recently remarried and living with her husband, her daughter, son-in-law, and granddaughter in a small two-bedroom duplex. While she was working at Clearvalley she was very active in the in-plant association, and she was vice-president at the time of the shut-down. She had experience in many areas of the factory. After losing the job she was unable to find employment and was spending most of her time caring for her ill daughter and her family. Rita and her family had felt the financial and emotional burdens of job loss, and she had experienced the loss of her own freedom and independence.

Joan and Bill Smith were a young married couple with a small child. Both worked at the plant for five years, Joan as a quality-control technician and Bill as a massager-operator. Both were devastated by the shut-down and had great difficulty trying to recover emotionally and financially. Eventually they had found new jobs that were either part-time or paid considerably less. They hoped that their situation would improve but were still worried about their future. Both were very critical of the company and the social-welfare system.

Throughout our interviews with these workers, one message was clear: financial problems persisted long after the shut-down. All of the workers had undergone financial strain, especially as it related to their family income.

Betty felt that her situation immediately after the shut-down was not too bad because her husband held a good-paying factory job. But she aptly pointed out the spin-off of her job loss and the entire plant closure: "It had a domino effect too. If you figure I have two kids and that meant if I lost my job then my babysitter lost her job.... It has that kind of effect. I guess all over too ... if one plant closes you have to figure that salesmen lost their jobs, and your box companies ... it is really a domino thing. And yeah, a lot of babysitters don't have kids any more."

Both Hanna and Connie went through economic adjustments, even though they still had some income to contribute to the household and their husbands were working. Hanna discussed how her change from a salary at Clearvalley to collecting UIC affected her life. She did not have the same economic flexibility as before, which resulted in less for herself and her children.

The loss of income, I mean I had to go to a much lower income. You had to watch yourself and watch what you were spending on that severance. Then you went to UIC, it is almost ... half of what you were making. So you didn't have that money to spend like you used to. You know you had to watch what you were spending.

For that whole year I guess up until I started school in September, I didn't have anything to spend on myself and that really bugged me.... Because you know you work so hard and you know that money that you make you deserve to buy stuff for yourself.... I was always so used to having money.... I used to buy a lot for the kids. Well, now you can't afford a lot of clothes and stuff for the kids like I used to.

Connie said she had to make adjustments, but they had some income so the strain was not too great. "As soon as I found out that I lost my job everything tightened up. No more spending money. I wouldn't even buy groceries, that's how bad it was.... And then after a while I got unemployment so it wasn't so bad.... We have been lucky so far because I've always had money coming in." Connie also talked about some of her family's new financial worries. "It wouldn't have been so bad except that a month before they told us we were closing down we just got a loan for a truck and it's not paid off until next year.... The house doesn't bother me, because we only pay 450 bucks a month for our mortgage. But it's the loan – our loan is 600 bucks a month."

Rita and her family experienced even worse conditions after the plant shut-down:

I always had money in my wallet. Well now I don't have money in my wallet.... We used to be able to save money and have all our bills paid on time, but now you can't do that. One might get paid and then the next month another one.... I don't like it because it doesn't give you very good credit standing.... It's not a nice feeling especially when you are obliged to pay your bills. You never used to shirk your bills. Like the other day I went into the bank and said to her [the bank manager], "Look, I'm not trying to shirk my responsibilities, I just don't have it." I said, "I know you don't want any excuses or anything so you're not getting any ... take a number with everybody else on my back." They [the bank] do not work it out, they just want money ... if you pay them a little bit then they call you the next day because they want more. Now my attitude is take a number and get to the back of the line and when I get it, you'll get it.... The bills are mounting up ... and I haven't had a vacation in sixteen years.

Rita will soon need a new car, and without one her future employment opportunities will be limited. "When you're getting to the desperate stage with your finances and everything, like we need a new vehicle now but I can't afford a payment. Once we are down to one vehicle then what are we going to do then? Where we are located you have to have two vehicles."

Bill and Joan's economic situation became desperate after their jobs were terminated. When Joan's UIC ran out, the entire family had to rely solely on her husband's UIC payments. They were constantly threatened with the worries of just surviving. Part of the frustration was in coming up against companies unsympathetic to their plight. One of those, according to Joan, was the hydro company. "When you get a knock at the door [and they ask] 'well, have you forgotten to pay your hydro bill?' 'Well no, we haven't forgotten to pay it.' Well they just don't care. 'Well I'm sorry, you've got five days to pay it or it's cut off!'"

Joan was even more impassioned when she described how her family lived on very little and how their situation became despairing.

Living on $1,000 a month isn't enough, especially when you have to pay a little more than half of that out just to rent. There is no money for hydro, and then we have a three-year-old son and he needs clothes, he needs food, medication, and we just couldn't get any help.... We even phoned our Member of Parliament for our area, and he wouldn't do anything for us. He told us his hands were tied. It's been pretty bad. A couple of times you really had to embarrass yourself going to a food bank. Nobody wants to know how degrading that is, when you have to walk in and say, can you help me? I was so embarrassed the first time I had gone ... I was so embarrassed, thinking I don't believe this. But with us having Tom [staying] in the living room and when you go to the cupboard and there's nothing there, you have to do something.

In addition to the economic impact there was the loss of freedom and independence, which involved both changing roles within the family and a change in living conditions. Some of the workers had difficulty adjusting to their new or lost roles within their families.

Bill had to accept the loss of his role as the major breadwinner. Both he and Joan had always worked during their married life, but he always enjoyed a higher income and thus an elevated status. Joan mentioned the difficulty her husband had with this change: "Then it was even harder when I found work and he didn't. It really bothered Bill knowing that I was going out to work and he was staying home with Tom." Bill's response to his wife's observation was, "Well, maybe I'm a little bit old-fashioned. But I believe that in order for us to make it nowadays both have to work. Joan's dad is from the old school and he thinks that the wife stays at home and the man is the breadwinner. Well, I was glad Joan finally got a job, and we finally got a little bit of breathing room. And then too, it's all her money, you know what I mean. All I had was my UIC, but her money was paying the bills and that was hard too because I get so bored sitting around doing nothing."

For Rita her job loss and unemployed status meant a loss of the independence and control that had taken her many years to achieve. It has been tough for her to comprise that independence.

I guess it's harder for me because I'm such an independent person and I don't like relying on anybody, I don't even like relying on my husband. I like that independence, it took me ten years to get that independence ... now I have to ask him for money. I raised two kids on nothing, I seemed to have more back then than I do right now....

It was very degrading having to depend on someone else after you've survived so long on your own. I mean I survived for ten years before I got remarried, with two kids to raise and I didn't have to ask anybody for anything. Now I'm married and my kids are grown up and I'm having to depend on somebody. It's mind boggling.

One of the adjustments, because of reduced finances, was a change in living conditions. Some of the group interviewed had to move to cheaper and more cramped housing. Rita and her husband had to have their children move back home because of finances and health. Her husband said, "Well, the kids have had to move in because I can't help them out, and that is partly due to [my daughter's] health." Bill and Joan could no longer afford to rent a house and had to move to a much smaller place, placing an extra strain on them. "First of all we had to move because we couldn't afford to stay where we were.... We went from a house to an apartment. This was hard because it was a big house and had a lot of room."

For a family a lack of sufficient space can make things difficult in the best of times. Rita talked about the lack of privacy at her house: "There isn't much room at my place to go and have privacy for anybody. It's hard on everybody." Bill was frustrated about being stuck at home with no personal space. "If you don't have anything to do and you're cooped up, you're bound to get on one another's nerves. You don't seem to have your space."

Corporate restructuring also brings a psychological cost – a broad category that not only captures the range of emotional consequences of job loss and unemployment but also includes five smaller areas: boredom, disruption of personal life routine, depression, strain on the family, and loss of contacts with friends. Hanna talked at length about the problem of not working and of how staying at home leads to feeling bored, which eventually turns into a rut.

For a while it is great [being unemployed] and then you get bored being around the house.... You kind of start going nuts because you don't have anything to do and you're stuck in the house.... But when the kids are in school and you're here by yourself, you get into a rut. You just get so bored ... you don't even feel like going outside of the house. It got that bad you just didn't feel like going out of the house. You were wishing you were working.... Being around the house when you were not used to it.... And I wasn't busy and you just get so lazy and then you just can't snap out of it, you know. I was always so busy at work and then you come home to not being busy.

Most of the workers we interviewed were women who had many other responsibilities apart from work both in the household and family and outside. When their daily routines and structures were disrupted by job loss, the rest of their lives suffered as well. With unemployment, school, or even a new part-time job, the women found it hard to re-establish their personal life routines. This ultimately impacted on their sense of control over their lives and their psychological well-being. Rita explained how she managed her personal responsibilities when working full-time, that she was able to extend herself and was very happy to do so.

Well, when I was working before I used to put my time between my mother, my kids, his parents, anybody else that needed me in a crunch or people who just wanted to talk, I would make myself available and everything. Now I don't answer the phone in case somebody wants me to do something or I don't want to hear about their problems.... I had myself where I was going twenty-four hours a day just going on four or maybe five hours' sleep a night. I was a much better person, I was a happier person. I made my time for myself when I got home, which was about an hour and a half before my husband got home.

Joan and her husband had an established routine that incorporated work, domestic responsibilities, and social life. But with job loss this routine was disrupted, and so too was Joan's self-worth. "When we worked at Clearvalley we knew that Sunday night we would go to bed at a certain time. We would get up and I would take Tom to the sitter and we went to work. Saturday and Sunday came along, Saturday was grocery shopping and we would be able to go out Saturday night. Sunday for me was cleaning and laundry day and all that. But after a while you get up in the morning and like, what for? What's the point of getting up? You've got nothing to do."

The former Clearvalley employees talked about the depression they felt as a result of their jobs being terminated. They spoke of their depression in different contexts, such as not being productive, not socializing, or desperation. When Connie lost her job she didn't know how to occupy her time. "Just the fact that I'm here all the time and I'm miserable. I just blow up at the kids.... I just get miserable sitting at home.... You can only clean your house so many times." Rita noticed that she became depressed when her life outside home was diminished. "And I don't have a social life any more because I can't afford that. This made me feel damn depressed, because I was one who used to go out all the time.... I like people and know a lot of people." Bill spoke of the feeling of helplessness when he lost his job and could not pick up the pieces. "Lately it's been the harder you fight the deeper the hole gets. Like you never seem to get out so you get into one big rut and it keeps getting deeper and deeper."

Some of the workers also noticed a strain on the family. Both Joan and Rita felt that their relationships with family members suffered as a result of their depression and apathy over their job losses. Rita revealed that her relationship with her husband had suffered at times. "You are constantly fighting, you don't want to be bothered. There are some nights when you think, don't even look at me and especially don't touch me." Joan and Bill fought more and felt enormous financial pressure, which strained their relationship. Joan worried about the impact of this on their three-year-old son. "And it's not fair on Tom because pressure gets to us ... even though we don't mean to, we yell more and you don't have as much patience. It affects him too, because when we are fighting over bills he kind of backs away.... When the pressure is so bad on us then we don't like doing the things we used to with Tom. You worry when you don't want to read a story book or go outside and throw snowballs."

The loss of contact with friends and co-workers at Clearvalley was another phenomenon the workers experienced. Some of them expressed this as a sense of loss and loneliness. Hanna missed the contact and felt this affected her motivation. "Well, you don't see those friends ... you just kind of lose contact.... The odd time you run into somebody, and for the first little while you keep in contact with your friends but then slowly you lose contact.... You really miss that and that was the biggest thing. You miss your friends, the people that you laughed with every day and joked with.... It was hard because you got lonely and you even get sick of cleaning the house.... I still had a car and I could just up and go, and go wherever. I wanted to, but I didn't." Rita also missed her friends. "There is a sense of loss ... there were people like my boss, that I really liked. We would get together and go bowling or something. But once the job ceased to exist everybody sort of went their own way and everybody is busy at different times.... It's not the same now. You spent most of your time with those people, and now you don't."

Some of the former Clearvalley employees passively accepted the plant shut-down and termination of their jobs. Others were not so passive about their circumstances. They were critical of the parent company and the consequences of the company's actions. They spoke about how they were treated, their confusion over why the plant had to be closed, and they were critical of what that company and others were doing and the impact on the community.

Connie was confused about the shut-down, especially since the plant was doing well. "It was so weird because they kept saying, oh we're doing so good now, Clearvalley will never close ... a week before [the shut-down] they were saying this to us and nobody knew." Jackie was also confused and believed the company could have saved the plant. "We can't figure out why they had to shut down. They could have fixed it up ... they could have done something. There were options, they could have worked without closing down."

Rita was angry with how Hillsdown treated the employees:

[They] didn't really care about anybody's feelings. They didn't treat you like you were human – like you were a number to be discarded and nobody had to worry about you once you were gone. They didn't know you and didn't have to feel sorry for you. You think a little more if you know the person and you have to terminate them, then you weigh all the pros and cons. These people, because they are sitting up in their big chairs, they didn't even have the decency to come and meet any of the employees to see what they are going to do. Like how devastating it was going to be. It was, we bought you this month and we're shutting you next month.... People just don't give a shit about other people.

Rita told a story that illustrates starkly why some employees see the management as insensitive to their plight. Some of the supervisors organized a farewell barbeque, inviting the management and employees. But the actions of the management left the employees bitter and offended.

They treated us as if we were nothing ... like we threw a big company BBQ as a farewell thing and we would invite management. Management was assigned to bring the food and this and that. Well, you know what a slap in the face is?... They rented a limo, a white stretch limo to bring these people to a country BBQ. They arrived several hours late. Do you know what people thought when they pulled into that farm?... They show up in a limo and all of us are unemployed. No, it did not go over well. Because people didn't think. What they [supervisors] were trying to do was give them [management] a gift. Well the gift should have been nothing to do with our party because some people walked away from that BBQ with snide remarks. It was just a bad taste. You do not drive up to a country BBQ in a white limo drinking champagne.

Some of the workers were critical of Hillsdown Holdings and other companies and the impact of their actions on the local community. Rita blamed the company and government for allowing plant closures. "You know I'm really down on government because they are allowing most of this. I just don't think big companies should be allowed to do that without thinking of the consequences.... I don't think they should have the right to buy small companies and shut them down for a tax write-off. As far as I'm concerned, that's what they do."

Joan and Bill expressed a criticism of both the company and others in their community. They witnessed their community being hurt because of some of the tactics of big companies. They used their own experiences as a departure point to examine what else was happening around them: "We also found out they are building this new Zehrs in [town] and apparently ... those people already working at [the old] Zehrs would have to reapply for positions. Then we found out that they are bringing people from [another town] to work there. And that's not fair, they should be letting the people here in [town] get the work. It's going to be in our community ... it's going to be us shopping there. It should be given to the people of [town] ... but they take work away from us."

When all is said and done, large companies have defended restructuring, arguing for the need to become more competitive in our current economic climate. Unfortunately, the fall-out from this tactic drops on the plant employees, who find themselves without work and struggling to survive. For these workers, a job is more than just a means to earn a living. Like everyone else, they form most of their relationships and develop their attitudes around a job. Their home, status, and place in the community, their friends and financial, cultural, and political associations are all determined by their jobs.

Blue-collar workers in general feel the impact of restructuring, but it seems that women, particularly older women, are hurt even more severely. Older women with little education will have difficulty finding comparable work. One of the Clearvalley workers was sympathetic and aware of the hardships these women face in finding jobs: "I'm sure that a lot of older women, you know the ones that are in their forties and fifties, it has to be hard for them.... I consider myself lucky because I'm younger and I have a chance to go back to school. But some of these women in their forties, fifties, and sixties ... who's going to hire somebody that old?"

The Dubious Direction of Our Food System

"Maple Leaf Foods has been rationalizing its businesses across Canada during the past two years in order to become more efficient operators in today's competitive environment." This rationale was offered in March 1992 by the management of Canada's largest food company to justify the termination of more than four hundred jobs at its Montreal meatpacking facility. In the preceding two years, thousands of people had been similarly affected in plant shut-downs orchestrated by this company across the country. In offering this rationale, the management of Maple Leaf Foods was by no means setting itself apart from other large corporate organizations. Similar rationales could, and indeed have, been used by numerous other firms to justify plant closures and relocation, usually to countries in places where temperatures are higher and wages are lower.

These rationales provide the "grease" for effectively reducing the friction between the often conflicting interests of the chief benefactors of transnational food corporations and the plant workers and primary producers who supply, operate, and maintain our industrial food system. Such rationales effectively slough off responsibility for the liquidation of our food industry by referring blame to a higher authority, usually the "impersonal" forces of the market economy. Because we are supposedly dealing with "natural" economic laws, opposition to corporate rationalization is said to be misdirected, if not completely futile. As one business journal tells us, "Darwin lives in business just as he lives in the jungle, and restructuring is an essential part of the process of renewal" (*Canadian Business Magazine*, November 1990:82).

Must we embrace the social and economic logic of this brave new Darwinian world with its credo of "adapt or die?" Certainly, the present directions of our economy are not in reality the result of the unfolding of "natural" economic laws. Rather, they are the result of conscious and planned strategies and policies orchestrated by powerful interest groups in our society and elsewhere as a specific response to technological and structural change. But that response is not the only one possible. Our fate is not in any necessary sense inevitable, although many Canadians may choose to believe it is so. What is needed, as an essential first step to recognizing alternative paths, is a broader, holistic, and historical evaluation of our present predicament.

CHAPTER NINE

CONCLUSION

• • • • • • • •

For several decades early in the 20th century, much of Canadian life centred around farming and farm communities. Moreover, the political struggles of agrarian producers figured prominently on the national agenda. In these struggles to achieve their objectives, the unity of the producers was an absolutely essential ingredient. When producers managed to hold together a movement exhibiting a strong unity of purpose, as they did on the Prairies, they were able to demand the attention of provincial and federal governments and often gain substantial concessions. The early organizational struggles also show the importance of establishing a solid political formation that expresses farm interests.

The experience of the agrarians in the political realm also provides lessons about the dangers inherent in alliances with the traditionally dominant Liberal and Conservative parties. In Canada, the failure of the agrarian movement to achieve some of its broader goals – restrictions on corporate concentration in the capitalist sector, more all-encompassing supply management legislation, a truly comprehensive co-operative movement, and, most importantly, its agitation for a more democratic political process – was to carry a high price for farm operators after mid-century. This failure made farmers substantially more vulnerable to the expansion of the capitalist agribusiness sector of the input suppliers, food processors, grain handlers, food retailers, and the like. With the metamorphosis of the CCF into a political movement that represented labour and urban concerns, the farm sector was left without a party to call its own, and most of the farm community gravitated back to the old-line parties. The policies enacted by the two older parties – to the extent that they touch on agriculture – have by and large been most beneficial to a narrow sector of producers, typically the largest producers with the most capital.

At one time, when the rural population formed a more significant proportion of the electorate, it was possible to imagine the Canadian farm community reviving an authentically agrarian party. Such a party has characterized the political scene in several of the advanced capitalist countries in the post-World War Two era, including Norway, Sweden, and France. In such countries agrarian parties have successfully pursued policies that support a broad range of farm operators, and in the process rural communities have been sustained, and, indeed, they have flourished. In Norway, for example, this political pressure has resulted in a series of comprehensive rural development initiatives that have made possible a relatively prosperous agriculture in the far north along the Atlantic coast, which in turn has nurtured human settlement in that zone. In Canada today, with their much reduced numbers, the best strategy for farm operators when it comes to protecting their interests is most likely one that works towards overcoming the numerous divisions among them. Presenting a relatively united front is the only effective method whereby the majority of producers can hope to lobby the party in power successfully.

The empirical evidence related to the history of the agribusiness complex shows the direct significance of corporate concentration in the processing sphere for farm operators, and for the rural communities they help to sustain. The development of the food-processing industry has provided at least some Canadian producers with a valuable market when other marketing opportunities were under stress or disappearing altogether, but it would be naive to ignore the form this development has taken and the mounting evidence that the integration of farmers into the processing sphere is essentially an asymmetrical relationship, with food manufacturers holding most of the cards. In some sectors of the food system this situation has existed for a long time, as the evidence of the 1937 Royal Commission on Price Spreads makes abundantly clear. The lack of balance in the distribution of power is especially acute in provinces where farmers have not successfully established marketing boards or other mechanisms that provide them with influence over contract conditions and the prices they receive for raw product. There are beneficiaries on the farm side, but in most commodity sectors of the food-processing and farming chain primary producers have been subject to a fairly high rate of attrition since the 1950s. This is true even in a sector such as dairy, which has one of the most comprehensive supply management arrangements. In addition to the loss of independence, increasingly only the largest and most heavily capitalized producers have been able to take solace in this system –

although the instability provided by recent trade liberalization policies has now brought into doubt the security of even these favoured farm operators.

The rise of the agribusiness complex brings into focus issues that are of concern not just to farm operators *per se*. The more general pressures in our market economy that promote corporate mergers, and the consequences of the astonishing absence of even the most minimal legislation that might serve to put a brake to this process, are readily apparent in our food system. The control of Canadian agribusiness is largely in the hands of very large and complex corporate empires. With this kind of transnational control, there is unlikely to be an equitable division of the food-system pie among the various Canadian sectors – including the smaller agribusiness firms, primary producers, and the consuming public.

Excessive corporate concentration produces business practices in the advertising realm that may seem rational from the point of view of the industrial firm, but are irrational and socially irresponsible from the point of view of Canadian society as a whole. While the food industry is not unique in this respect, its heavy dependence on advertising closely connects to what is a wider flaw in how our society is organized.

Yet another salient issue is the relationship between a high degree of concentration of economic power, as we see in the food system, and the continuing viability of a democratic politics in this country. Political philosophers going back at least as far as Jean-Jacques Rousseau have warned about how the excessive polarization of wealth in society can be ultimately corrosive of the political processes that seek to safeguard democratic participation and decision-making. There is no doubt that in the food system, at least, we are witnessing the accelerated polarization of resources, both within the farming community and among the various players in the agro-food complex.

Where Is Our Food System Headed – and Do We Want to Be There?

It is still too early to formulate final conclusions about the current "restructuring" that is transforming Canada's food economy. Nevertheless, it is not too early to assess the fall-out of the process so far. The food system has been particularly hard hit by the assemblage of policies promoted by the new-right agenda in Canada. Food-industry plant workers have been especially ravaged by restructuring, but farm operators have not, and will not, emerge unscathed. What can we say about this experience so far? With the wolves gathering at our collective door, could it be that our longstanding inattention to the conse-

quences of corporate concentration and the tremendous power this has conferred on a very small element of our society have played a key role in rendering us so seemingly helpless to protect ourselves? The "adapt or die" credo of
today's business world must be appreciated for what it is – a self-serving
ideology designed to disarm the forces that could oppose the interests of
transnational firms and the people who own them.

We need to review a few key developments that characterize our present.
Our food system is increasingly being captured by transnational businesses
that are consolidating regionally or continentally integrated business operations as part of their overall strategy to maximize profits worldwide. The national origins of capital are increasingly becoming disassociated from the place
of production, as firms seek to find materials and process raw product in parts
of the globe that offer special cost advantages. For much of the last decade,
national governments have been adjusting their regulatory mechanisms to
accommodate transnational interests. For the most part, this has meant an
overwhelming tendency for state withdrawal from the economic sphere – a
conscious and planned liquidation of the state's once vigorous role in this
realm.

In this context – an ascendant private sector led by transnational corporations and the waning force of the state and other institutions such as organized
labour that have historically balanced the "private" and "public" spheres of
interest – it is clear what the outcome of the "adapt or die" corporate credo
will be. Those few players *with* substantial economic clout will adapt, with
varying degrees of success. Those players *without* such resources, like the
Clearvalley Farm plant workers, will be left to fend for themselves without
the educational and social service infrastructure that might have made some
kind of adaptation possible. Indeed, it is likely that most Canadians will be
hurt by a continuation of the drift towards reducing the role of the state in
society and leaving the country at the mercy of highly mobile, stateless corporations that are accountable to no one but their owners. As economist Kimon
Valaskakis argues, if this continues, "What is known as the 'public interest'
would be entirely drowned in a sea of private sector and selfish interests.
Quite probably, many factors of production would migrate away from Canada,
leaving the country 'nice and quiet, clean and green,' but depopulated and
poor" (*The Globe and Mail*, Oct. 31, 1992:B4).

Today, in our headlong rush to privatize, deregulate, and remove trading
barriers, much is being sacrificed on the alter of higher economic growth. We
are expected to tolerate the tremendous social disruption of corporate restruc-

turing and rationalization, presumably because, as we are told, it will result in more efficient businesses, with higher overall productivity. The attack on Canada's supply management system in agriculture is premised on much the same argument: dismantle supply management and revert to private sector arrangements and, it is said, the market will increase efficiencies and help to further stimulate economic growth.

In times of deep economic recession and social distress, the goal of economic growth can seem extraordinarily alluring. But in a globalized economic environment, to equate economic growth with the automatic betterment of the lot of the bulk of those who constitute our food system – primary producers, plant workers, and clerical and sales personnel – is a tragic mistake. It is possible – and highly likely under current conditions – that higher levels of economic productivity will lead to *lower* levels of well-being for substantial numbers of the population. Indeed, this is precisely what has happened in the United States. Considerable gains in manufacturing productivity were made there in the 1980s, a decade of massive deregulation and corporate restructuring, with movement towards the Southern states where lower wages and laxer environmental regulations apply. Yet an estimated 40 to 70 per cent of the increase in income went to the richest 1 per cent of all families.[1] At the same time, lower-middle income and poor families in the United States experienced a considerable erosion of incomes throughout the decade. This evidence describes nothing less than a growing polarization in U.S. society. In Canada, despite our much-lauded social programs, there is increasing evidence that the same process is at work. A recent study in Toronto found that the bottom 50 per cent of income earners had experienced a decline in real incomes in the years after the mid-1980s, with the bottom 10 per cent of income earners facing a decline of some 50 per cent. Other studies have similarly documented the dramatic erosion of family incomes throughout Canada during the early part of the decade, and in May 1993 Statistics Canada reported that family incomes for 1991 had dropped down to their 1976 level (*The Globe and Mail*, May 4, 1993).[2]

As a society we cannot expect to benefit from the increases in efficiency and productivity achieved by corporations operating in our economy if we acquiesce in the dismantling of the mechanisms that have ensured that at least some of the gains were actually spread around, and not captured by just a few. In fact, with the continuation of our head-long rush to embrace market mechanisms as the final arbiter of "who gets what," we cannot even be sure that transnational corporations will be around at all as productive entities in the Canadian economy.[3] If they are, it will most likely be at the price of "harmonizing" our society – from health care to welfare – to a lower level. If we continue down our present path the only question is, how much lower can we go? The current realities of the United States and Mexico suggest the spectrum within which this question will be answered.

Paradoxically, it may well be that real progress in our food system, and in the wider society as well, depends upon the strengthening of the mechanisms that in the past squeezed a degree of equity out of economic growth – or upon the elaboration of new mechanisms that will do this job in a changed economic environment. An equitable income distribution need not be seen as a deterrent to economic progress. On the contrary, in recent decades the advanced capitalist countries that have achieved the most equitable income distribution have also achieved the highest increases in productivity and general economic growth (Schlefer, 1992:118).

In the food system, as elsewhere in Canadian society, the possibility of achieving a more equitable path of development and the social stability that only greater equity can secure requires a successful challenge of the powerful interests that have captured the economic and political agenda. It requires overcoming divisions that have for many years fractured the farm community and set one commodity group against the next. It requires recognition of the necessity of setting aside sectoral interests for the greater good, and recognition that power can only be achieved through strategic alliances between primary producers and the labour movement, which today is one of the most unified elements in the Canadian struggle to defeat the new-right agenda. It requires building new alliances with natural allies elsewhere in society and building an understanding of the importance of nurturing an agricultural economy on Canadian soil, alongside a food industry that responds to domestically established priorities for food quality and environmental standards.

Unless these challenges are met, most of the players who make up our food system will remain weak and ineffectual. They will not be able to forge a viable alternative to the current direction and its central organizing principle – "the weak to the wall, and the strong shall take all."

GLOSSARY

• • • • • • • •

Co-packing This is an arrangement whereby one processing firm packs product for another firm, usually under the other firm's brand name.

Economies of scale These are said to exist when an expansion in the scale of production of a firm or industry results in an increase in output that is proportionately greater than the increase in costs.

Market share This usually refers to the amount of the sale of the product or products of a particular company as a proportion of the total sales of the product or products of the industry as a whole.

Predatory pricing policies This is typically a practice of a corporation with considerable market power in an industry. This power is used to set prices at levels that smaller firms with fewer resources cannot match for any length of time, so the smaller companies are therefore forced out of business. Such prices can also discourage the entry of new firms into a market.

Product differentiation A product is considered to be differentiated to the degree to which consumers consider alternative products to be imperfect substitutes. Advertising is essential to this process. When a product is highly differentiated the firm producing it enjoys a degree of insulation from the competition of rival producers. The condition also greatly increases the pressure on retailers to stock the brand.

Social differentiation This term is used to describe the process of decomposition of a family or peasant farm economy as a result of the working of market forces and powerful external economic actors.

Value added This refers to the value added to the raw materials and services purchased by a firm, through the processessing or production activities of the firm.

NOTES

• • • • • • • •

Introduction: Production, Consumption, and Power in the Food Economy

1. For example, the main objective of agricultural policy in Sweden throughout most of the post-World War II period was to safeguard the provision of food during times of peace, blockade, and war. Food security, then, has been closely tied to a national defence policy that established the level of support that would go to agriculture to ensure the agricultural reserves deemed necessary to meet a sustained national emergency (see OECD, 1988:21-22).

2. These figures are taken from the "Preliminary Findings: Community Food Needs Assessment" undertaken by the Edmonton Food Policy Council in 1991. For the survey, hunger was defined on the basis of answers to a five-question scale about how well people were meeting their food needs.

3. The implications of the expansion of capitalist social relations in society are discussed by Marx in some detail in "Results of the Immediate Process of Production," published as an appendix to volume one of his *Capital* (1976). For the debate on the reasons underlying the persistence of the family farm enterprise, see Friedmann, 1978, 1980; de Janvry, 1980; Johnson, 1981; Goodman and Redclift, 1985; Lianos, 1984; Mann and Dickinson, 1980.

4. Terms such as "agribusiness" (Goldberg, 1957), "l'économie agro-alimentaire" (Malassis, 1973), "agro-food systems" (Domike, 1976), and "agro-food chains" (Arroyo et al., 1981) have been used in attempts to reconceptualize the complex phenomena encompassing food production, transformation, and distribution today.

1 Building Solidarity: The Early Farm Organizations

1. Towards the end of the 19th century there was a strong tendency for the amalgamation of enterprises by a few increasingly powerful capitalists in many sectors of the economy. The large firms created by the swallowing up of many smaller enterprises were originally called *trusts*. This tendency, and the grow-

ing preference of the most powerful capitalists in each sector of the economy to collude in the setting of prices and other restrictive practices, eventually brought a strong reaction from other sectors of society. In the United States anti-trust legislation was enacted to prevent mergers and acquisitions that might lessen competition and create market restrictions that were considered to be to the detriment of smaller entrepreneurs and the consuming public. In Canada such legislation has historically been very weak and ineffective in preventing the development of a high degree of corporate concentration.

2. For a discussion of the restrictive trade practices of the elevator companies by a contemporary observer, see the comments of R.L. Richardson cited in Wilson (1978:26-27).

3. From *Report and Evidence of the Royal Commission on the Shipment and Transportation of Grain* (1900), pp.8-12 (cited in Wilson, 1973: 29-30).

4. For evidence of the dramatic price fluctuations of wheat, see the data provided by Lipset (1968:46), who notes that in the early decades of this century wheat prices in one year could decline to less than half of the previous year's price.

5. The situation was somewhat different for Eastern farmers, who practised a mixed farming regime and were as a result much less vulnerable to the price fluctuations of any one farm commodity.

6. The commission's *Report* gives valuable insights into how Prairie capitalism actually worked at that time. It contended that in the country's largest grain market, the Winnipeg Grain Exchange, "The larger milling and elevator companies operating elevators in the country, and owning and controlling most of the terminals as well, have overwhelming advantages over all the other members (i.e. commission merchants and track buyers).... They can buy large quantities of street grain cheap, they can enhance their profits by malpractice in both initial and terminal elevators, they have the income derived from both the storing and handling of the grain, and they can obtain special privileges in transportation and banking. Because of these advantages they exercise a controlling influence in the Exchange.... It is evident that the grain business of Western Canada is in the hands of a powerful monopoly, in which a few large milling companies are supreme and the large elevator companies hold the second and the only other place" (quoted in MacIntosh, 1924:47).

7. This letter is cited in an advertisement by the United Grain Growers Co-operative, an organization that emerged from Partridge's original Grain Growers' Grain Company (*Country Guide*, September 1991).

8. The major errors included overpayments for existing elevators, the purchase of elevators that were already losing business, and placing restrictions to limit the government elevator system to the less profitable storage of grain, when private competitors could engage in the more lucrative grain-buying business (MacIntosh, 1924:40-41).

9. The precedent for such a system had been set by successful marketing boards established during the First World War.

10. Wilson (1978: ch. 10) has more detail on the pooling concept and its history.

2 The Role of Farmers' Parties: A New Form of Power

1. This success also undoubtedly had something to do with the fact that women had won the franchise during the war. The Conservatives decided to hold a referendum on the wartime prohibition law. This gave the farmers' party an edge, as it was staunchly prohibitionist and therefore particularly attractive to the newly voting rural women who tended to be fervent followers of the temperance movement.

2. By 1923 the farm and labour groups were unable to agree on common candidates in some ridings, resulting in considerable hostility. For a report of such a situation that developed in Guelph, see *The Globe*, April 23, 1923.

3. The views of each side in this debate were given much play in the pages of the *Farmers' Sun*. See especially, August 19, 1922, "Premier Defines His Stand on Broadening Out Policy," p.1, and "United Farmers Will Stand by Principles Says Morrison," p.1. See also August 22, 1922, "Party System Now Obsolete Says Morrison," p.1; and August 29, 1922, "Can Be Harmony in New System, Says Morrison," p.1. An editorial on the dangers of being absorbed by the Liberal Party should broadening out be pursued was printed in the *Farmers' Sun* on September 2, 1922.

4. Another prominent agrarian leader opposed to much of Drury's strategy was W.C. Good. See his address to the 1924 convention of the UFO (Good, 1958:169).

5. This point was discussed in an editorial by J.J. Morrison in the *Farmers' Sun* some years later (*Farmers' Sun*, February 4, 1932).

6. As Janine Brodie (1990:97) argues, Vernon Fowke has provided one of the most influential and inclusive interpretations of Canada's National Policy. She notes that according to Fowke, "The National Policy was the first in the series of overarching development strategies adopted by the Canadian state, the central objective of which was to 'transform the British North American territories of the mid-nineteenth century into a political and economic unit.' Confederation was its major constitutional instrument, while its other dimensions, 'western land and settlement, the Pacific railway design, and the system of protective tariffs, took shape in the decades after 1867.'"

   Other major items on this "platform" included nationalization of the railways and telegraphs, along with a progressive tax on incomes and a tax on inheritance and excess profits and the federal enfranchisement of women who were already accorded the franchise in their province (Wood, 1975:346).

7. For an excellent discussion of the links of the agrarian leadership with the Liberal Party in the West, see Morton (1950), especially chapters 2 and 3.

8. Morton writes: "The Union Government was, unintentionally, a powerful demonstration of non-partisanship. The electorate of the West came out of the election of 1917 purged of old party loyalties.... The effect on the old parties in the West was disastrous. Their federal organizations were destroyed. The Conservative Party lost its identity as a federal party.... Within the Liberal Party there was civil war and the factions smarted with bitter animosity" (1950:60).

9. For more details about the life and political achievements of this outstanding representative of the farm movement, see the fine in-depth study by Terry Crowley (1990).

10. Morton (1950:292) argues that the Progressives did have a positive impact on this plane. Because of their influence, "The old order of Macdonald and Laurier, when party affiliation was hereditary and party chieftains were almost deified, was not restored. The two-party system did not return in its former strength. The rules and conventions of Parliament made provision for more than two parties. The electorate became more independent ... the authority of the whip became lighter, the bonds of caucus weaker, than in the old days."

3 The CCF: A Lasting Political Vehicle

1. The debate confronted the issue about non-pool farmers receiving the benefits of the pooling system in the form of stable prices but paying no costs, and how this group could undermine the pool by pursuing their private interests when the pool's services were not as attractive (see Fairbairn, 1984:104-6).

2. This closure was orchestrated by long-time grain trader John I. McFarland, who, as a close friend of new Prime Minister R.B. Bennett, was strongly pushed by the banks holding the Saskatchewan Wheat Pool's debt to become the pool's general manager. While the pool leadership did not favour this, impending bankruptcy gave them little choice (see Fairbairn, 1984:101-2).

3. These connections are discussed in some detail in Finkel (1979), chapter 2.

4. Courville (1985) for one is sceptical about the extent of this influence, arguing that many members of the leadership of the provincial Progressive Party were quite conservative in their orientation and returned to the fold of one or another of the old-line parties once the Progressive cause began to wane. Furthermore, he claims that radical elements that fed into the CCF – such as the Farmers' Union – were not really part of the heritage of the Progressive movement. These criticisms seem to miss the point, however. Whatever its political hue, the Progressive Party established a precedent for a third party as a legitimate means to pursue political objectives outside the old-line parties.

5. For the full details of the CCF's rise to power, see, for example, Young (1969); Lipset (1968); Silverstein (1968); and McNaught (1959).

6. There were agrarian elements participating in the process, principally from

Ontario and Alberta, who were wary about the socialistic tenor of the manifesto and were less than enthusiastic about all efforts being put into creating a
new political party to foment social change (see Young, 1969:48). At the other
extreme, members of the B.C. labour movement, more closely under the influence of Marxian socialism, were not content with the document's program
for achieving change by legislative means; they were more inclined to view
the institutions of Parliament as instruments controlled by the dominant class
and unlikely to serve working-class interests.

7. All quotations come from the *Cooperative Commonwealth Federation Programme*,
   adopted at the first national convention held at Regina, Saskatchewan, July
   1933 and reprinted in McNaught (1959:Appendix).
8. The Beveridge plan was a comprehensive report to the British government on
   social insurance. It was, for its time, far-sighted and innovative. It received international recognition and had an impact on the thinking of policy-makers in
   many countries in the following years.
9. The experience of co-operative farming at this time is reported in Henry
   Cooperstock, "Prior Socialization and Cooperative Farming," in Blishen
   (1964:2-7).
10. Compulsory check-off is a principle usually associated with the struggles of
    the urban labour movement. It holds that dues to support the union or em-
    ployees association are to be paid by all employees within an establishment,
    whether they are members of the employees union/association or not, on the
    principle that all of them benefit from the gains made by the union.
11. See Lipset for electoral data for 1934 and 1944 (1968:200, 206, Tables 15, 18).
12. For a survey of press attitudes both within and outside the province around
    this issue, and the doctors' strike in particular, see Taylor (1978:307-13).
13. It apparently had an important influence on the then-sitting Royal Commis-
    sion on Health Services, chaired by Chief Justice Emmett Hall. The Commis-
    sion's recommendations gave medical care insurance the highest priority for
    action by the Canadian government (Taylor, 1978:328).

4 The Institutionalization of Agrarian Power
1. See his letter to Prime Minister Bennett, cited in Wilson (1978:460).
2. Producers in some European countries, with existing tariffs and quotas, were
   getting in the range of $2.00 a bushel at the time (Finkel, 1979:74).
3. Costs for these producers were also higher because of the higher standards
   they had to maintain and the extra expenses incurred for refrigerated transportation.
4. See the editorial featured in *The Ontario Milk Producer*, "Complete Organization of Dairy Interests Needed" (April 1932:335).
5. The quota system has been a crucial part of the regulatory apparatus that

brought order to the milk industry. In educating the producer about the value of a quota system, *The Ontario Milk Producer* stated: "The quota presupposes a surplus of milk in the available market. If there were no surplus of milk, there would be no need for a quota. Of the total quantity of milk available in the market, the quota is assigned pro-portion of that quantity which the distributors estimate they can use in bottling. Formerly, we had no control over the surplus, and much of it came into the market in spite of holdbacks, and gave to any dealers who chose to buy it the advantage of cheap milk. By the quota each producer is expected to ... keep his surplus at home, or if he prefers, ship it in to ... his own dairy.... But in fixing the quota to their shoppers, the distributors must allow for some surplus to take care of upward variations in demand, and to allow for some shippers falling short of their quotas" (November, 1933:35).

6. Finkel also argues that the capitalist dairy processing sector, or at least the more established element of it, had a vested interest in securing more stable market conditions (1979:46).

7. The powers of the Board are discussed in *The Ontario Milk Producer*, "Some Questions That Are Being Asked About the Milk Control Board and the Answers" (July 1934:28).

8. That a stable price structure was still rather precarious was alluded to in an address by the president of the Toronto Milk Producers Association in 1936 (see *The Ontario Milk Producer*, 1936:154).

9. This discussion is based on an interview with George McLaughlin, first chairperson of the Ontario Milk Marketing Board (March 6, 1992).

10. The salience of these factors was suggested by a reading of Biggs (1990), and in the interview with George McLaughlin. McLaughlin also noted that smaller fluid-milk processors were often much more resistant to the idea of a pooling scheme for milk than other players.

5 Corporate Concentration in the Early 20th Century

1. These stages are summarized in Malassis and Padilla (1986:10). My discussion of these stages draws heavily from this work, unless otherwise indicated. See also Malassis (1973, 1986).

2. The primary processing activities were flour milling, distilling and brewing, vegetable canning, and cheese-making. However, both the inputs for the processing activity and the demand for its products were localized, geographically speaking. Typically, the early processing companies made one product in a single plant (Connor, 1980:6).

3. This is reported in Chapter IV of the Royal Commission Report (Canada, 1937:47). To have some idea of what $1 million in output signified at the time, it is useful to know that the average annual manufacturing wage in 1930 was

only about $1,000. Factory workers in the canning industry were paid about twenty-five cents an hour (Canada, 1937:364).

4. For a discussion of the three main periods of technological change in Canadian agriculture, see Winson (1985:424-25).

6 Consolidation of the Agro-Food Complex

1. That is, these units are not capitalist in terms of the relations of production on the farm, because of the ability of producers to substitute capital for labour to such a high degree in agriculture.

2. These are subsidiaries of a larger firm, and would include the agricultural implement division of major automobile manufacturers such as Ford Motor Company and the agricultural chemical division of multinational chemical companies such as Monsanto.

3. Burns notes, "The strength of restrictive practices legislation after 1956, relative to the government's soft approach on monopolies and mergers may well have promoted amalgamation, rather than collusion, as a defensive tactic against competitive pressures and declining market share against a background of slow market growth and excess production capacity" (1983:3).

4. Calculated from OECD (1979:257), Table 81, "The World's 100 Largest Food Processing Firms in 1974." Only twenty-seven of the one hundred largest food-processing firms had no foreign subsidiaries.

5. The expansion of the tobacco giant Philip Morris into the U.S. brewing industry when it bought the Miller Brewing Company was fuelled by an all-out sustained advertising campaign. Revenues gained from its lucrative tobacco business were pumped into its new brewing company. Advertising expenditures for Philip Morris-Miller increased by 1,000 per cent between 1970 and 1982, to the point where it spent more on advertising than the combined total of all other brewers except the Anheuser-Busch Company (Connor et al., 1985:255).

6. This is documented in Statistics Canada's *Inter-Corporate Ownership* directory for 1992 (cited in *The Gazette* [Montreal], June 6, 1992).

7. Ault reportedly took over more than fifty independent dairies, giving it a revenue in 1992 of $1.2 billion. In Ontario it owned such well-known brands as Sealtest, Silverwood, and Borden Dairies, Black Diamond cheese, Lactantia butter and margarine, and Haagen-Dazs ice cream (*The Globe and Mail*, Sept. 11, 1992).

8. Scherer notes evidence presented by the Procter & Gamble firm on this score (see 1982:198). See also Schmalensee (1978).

9. This is in keeping with an established theorem in economics – the Dorfman-Steiner theorem – that argues that if sales rise with increasing advertising expenditures, more will be spent on advertising the higher the price-cost mar-

gins are; the price-cost margins being defined as sales minus material and labour costs, divided by total sales (see Scherer, 1982:212).

10. Interview with Susan Cox, Assistant Executive Director, Daily Bread Food Bank, July 29, 1992.

11. The Child Care Resource and Research Unit at the University of Toronto estimates the annual costs (in 1993) of operating a child-care facility at about $6,500 per child. This figure estimates the average costs for caring for infants, toddlers, and school-aged children combined (communication with the author, March 1993).

12. Details on the financial pressures forcing people to use food banks was gathered in a survey conducted by the Daily Bread Food Bank. Calculation of the number of beneficiaries of the transit subsidy used the value of a monthly pass on the Toronto Transit Commission, which was $65 in July 1992, as a base for calculating transit costs. Fifty per cent of the yearly cost of $780 is $390. For $100 million this subsidy could be provided to 256,410 people.

13. A *Globe and Mail* report on the foreign takeover of the company notes that reporters were expressly excluded from the stockholders' meeting that was to vote on the move, and that *The Globe and Mail* had to buy stock in the company just to gain access to the meeting where the fateful decision was made (October 26, 1956).

14. Interview, June 1989.

15. Interview with a member of the provincial association of food processors, July 1987.

16. Interview, June 1989.

17. Interview with a member of the provincial association of food processors, July 1987.

18. Interview, June 1989. This discussion of the impact of the predatory pricing policies of the dominant firms in this industry was also corroborated in an interview with another long-time processor from the Niagara district (July 1987). This individual noted that during the mid-1980s, "An international firm decided they were going to expand their market portion of whole pack tomatoes. They decided to do that by putting almost a loss-leader deal on tomatoes, so in consequence the whole group of us suffered because the other large [firm] that had a large portion of the Canadian market said, hey, this new big guy coming along is not going to take a portion of the market we already had.... Sure, they hurt each other, but these guys have the wherewithal to hurt each other, but the others are killed.... Another year like the last couple of years, we just won't be around in whole pack tomatoes."

19. Quality and uniformity of supply involve such factors as produce size and shape to facilitate market preferences, and mechanical handling, colour, and texture, which can determine the suitability of produce for new production

techniques, such as rehydration. Bacteriological quality is especially important in the case of dairy produce (OECD, 1979:145-46).

20. Interview, January 1988.

21. Interview, January 1988.

22. This study was first reported in Winson (1988).

23. Details from production contracts obtained by the author. In the case of a dispute over quality, provision is made that a government inspector "may" be called in "for clarification of standards."

24. Interviews with processors and farmers supplying processors, June-July 1986.

25. Interview, July 1986. It seems that the rigid specifications demanded by such large-volume buyers as fast-food chicken outlets have so far discouraged the introduction of labour-displacing technology such as modern cut-up machinery, at least in the plants that supply these customers. Until the poultry coming into the processing plant is more standardized, human labour will still be required to cut the birds up to the exact proportions required by the fast-food chains. The poultry plant in our study with the new cut-up machinery did not have a contract with the fast-food retail outlet.

26. For the apple farmers, the possibility of selling their crop on the fresh market did not appear to be a viable alternative, mainly because of the limited development of local marketing channels for such commodities in the province. Interviews with farm operators, June-July 1986.

27. Prices and contract conditions have been more attractive for potato farmers than for other vegetable producers in Nova Scotia, a situation indicated by the relatively few complaints coming from potato producers in our discussions with them. This situation may be a reflection of the processor's view that small-size farm operations are able to produce a higher quality crop than the processor's own farming operations have been able to do. Interviews with the potato processor and farmer/suppliers, June-July 1986.

28. This and the other conditions are specified in the "Agreements" negotiated for such crops as peas, beans, carrots, and cauliflower under sections dealing with "contract provisions" (Ontario Vegetable Growers' Marketing Board, 1986a,b).

29. This situation is termed "by-passed" acreage and covers acreage that "at one point is suitable for harvesting and suitable for processing and is not harvested because of an insured peril" (OVGMB, 1986a).

30. The Warnock Hersey study, for instance, found a six-cent a pound difference in favour of Ontario pea producers in 1966, and a difference of five-cents a pound for beans, again in favour of Ontario producers.

31. This may help explain why the Consultative Task Force on the Canadian Food and Beverage Industry (Wygant, 1978), a body that was heavily dominated by representatives from the nation's largest food processing companies, expressed consistently negative views on marketing boards.

32. More details of this study can be found in Winson (1990). Only processing firms directly involved in agricultural production and/or with farm operators were included in the study. Due to several periods of corporate merger activity, of the approximately two hundred relevant Ontario firms in operation in this subsector by 1950, only forty existed in 1987, when this study was initiated. Of those forty firms, I was able to obtain data on twenty-six of them, including all of the large-size and most of the medium-size companies.

33. Although this processing firm still contracted with many small growers, it had rationalized its procurement practices considerably over the years, so that it procured its crop from one-quarter the number of growers that supplied it twenty years earlier.

34. Gamma and Somer's d were also computed to provide a quantitative validation of this apparent relationship. Their values are .857 and .654 respectively, indicating a moderately strong relationship. Gamma and Somer's d are measures of association for ordinal data based on the logic of pair-by-pair comparison, with values varying from -1 to +1. Somer's d is a somewhat more conservative measure that incorporates tied ranks on the dependent variable. Both assume that variables are linearly related.

35. See Winson (1990:387-89) for more details on this supporting evidence.

36. However, at least one medium-size processor, and several small ones, felt that it was *small* farmers who took more care in their farming practices. By "efficient" the large processors may have been referring more to the administrative capacities of large farmers, rather than their farming practices. Our research did not clarify this matter, unfortunately.

37. Interview, July and October 1987.

38. Interview, July 1987.

39. Interview, August 1987.

40. Interview, July 1986.

41. Interview, November 1983.

42. The farm-product price index had actually declined substantially in the mid-1980s. By 1990 it had recovered to the level existing in 1981 (Ferguson, 1992:3).

7 Food Retailers: The New Masters of the Food System

1. This refers to the *Census of Merchandising and Service Establishments* (Canada, 1933), and in particular to the section titled "Food Chains in Canada:1930."

2. Information on the "operating expense ratios" for different size stores is reported in the *Census of Merchandising and Service Establishments, 1931*, p.4.

3. Data taken from DBS, *Census of Merchandising and Service Establishments*, "Food Chains in Canada:1930," Ottawa, 1933.

4. Chain store operations are defined in the U.S. census as operations with more than eleven stores each (see Marion, 1986:294-98).

5. The 1,800 threshold that triggers a Department of Justice challenge to the merger is based on a four-firm ratio, it should be noted. A threshold based on five firms would be somewhat higher. Nevertheless, the Canadian index is so high that it would still substantially exceed a revised threshold based on the top five corporations.

6. Marion (1986:302) notes that court testimony and trade literature confirm that the strategic plans of grocery chains are usually formulated at the city or metropolitan level.

7. It is not surprising, then, that one well-known observer of the business community in Canada has noted that Calgarians pay more for their groceries than do the citizens of any other major centre in the country. Tiggert, *The Edmonton Journal*, May 26, 1987.

8. The allies at universities have typically been agricultural economists firmly committed to the free-market nostrums of supply-side economics. Curiously, with a few notable exceptions, their concern with impediments to the working of market forces has been almost entirely focused on the role of producer marketing boards. They have shown little interest in investigating the problems for competition that might be posed by the dominance of much of the Canadian food industry by a remarkably few very large conglomerate food corporations whose economic power has been enhanced through a high degree of economic power via vertical and horizontal integration. The tenor of their pronouncements as reported in the press makes it seem likely that many of them are not aware of this situation.

9. This is indicated by data provided in *The Globe and Mail Report on Business Magazine*, "Top 1000" (1991) and calculated by Ferguson (1992:41).

10. See "Profits Fall 16% in Quarter," *The Globe and Mail*, Nov. 2, 1992:B1. While profits for the distribution sector as a whole were down 73 per cent, they were up 17 per cent for food retailers.

11. Weston's subsequent acquisitions in the distribution business included the Alberta-based Jenkins Groceterias in 1959, OK Economy Stores in Saskatchewan, the wholesaler Kelly, Douglas & Co. in British Columbia, Atlantic Wholesalers in the Maritime Provinces in 1977, Star Supermarkets in 1982, and the U.S. food wholesalers Hickman, Coward and Wattles Inc. and Kotok and Heims Corporation in 1986. Also in 1986 Weston acquired the Halifax-based Capital Stores retail chain and twenty-six retail food outlets owned by the Kroger Co. in St. Louis, Missouri. In 1987 it purchased the Mr. Grocer franchise business of the Dominion Group, and more recently it acquired the fourteen stores owned by Fortino's Supermarkets in Ontario. In the Western provinces Weston controls eighteen superstores and twenty-eight supermarkets through its Westfair Food subsidiary. *Financial Post* Yellow Cards (1990) and *Canadian Grocer* (August 1991).

12. As one Borden executive noted a few years ago, "Making private label is a cancer. The better you do it the worse things get, because you erode your own brand's share" (noted in Connors et al., 1985:223). According to Bitton (1985:69-70), in Canada private label products accounted for 20 per cent of total retail sales in 1982, while generic products accounted for about 5 per cent of sales.

13. Interviews with Ontario food processors in 1987-88.

14. See the *Annual Report for Loblaw Companies Limited*, 1991, p.6.

15. See ibid., 24.

16. See *Royal Commission of Inquiry into Discounts and Allowances in the Food Industry in Ontario*, Judge W.W. Leach, 1980.

17. See ibid., 24.

18. For more discussion of the pressure of food retailers on food processors, see OECD (1979:44).

19. This is noted in the report of the Harvard Business School Club of Toronto's "Report to the Daily Bread Food Bank on Marketing Advice to Increase Food Flow from the Manufacturing and Distributing Sectors of the Food Industry," December, 1991, mimeo.

8 Restructuring and Crisis in the Canadian Food System

1. He did this, for instance, on the TV program "Freedom to Choose," broadcast on PBS in 1983.

2. See "Food Processing Closures in Ontario as of October 31, 1991" (Heisler, 1991).

3. Much of the information here comes from a study of food industry restructuring begun in November 1992 and from the local newspaper's coverage of the plant closure. To explore the human cost of restructuring, we conducted interviews with a representative sample of about twenty employees, whose experiences of job loss and trying to cope in the long term illustrate the impact on the immediate community, family life, and the wider rural community of a corporate reorganization over which they had no control.

   We used loosely structured, open-ended questionnaires. In order to preserve the anonymity of our respondents we have changed the names of the firm and the respondents. In this section we use the responses of six subjects who offer a good range of situations and experiences. These six people generously shared their time and experiences with us. We have tried to preserve the authenticity of their stories, experiences, and responses.

   In organizing the information from the interviews we selected themes that turned up throughout them. The plant closure and subsequent employment termination were experienced by plant workers in various ways. We have chosen to categorize these as: the economic impact; the psychological cost; loss of freedom and independence; and the expression of a critical consciousness about what happened.

9 Conclusion

1. This is reported in Schlefer (1992), which notes that the variation in the percentage of the income increase depends upon whether one adjusts for the declining size of families.

2. The data for Toronto was in a study by the Social Planning Council of Metro Toronto, presented to the Conference "Canadian Political Economy in Hard Times," University of Toronto, Jan. 22, 1993. For a study that deals with the erosion of family incomes across Canada for the earlier part of the decade, see Duffy and Pupo (1992:27).

3. The decline of Canadian manufacturing since trade liberalization was inaugurated as a national priority through the Free Trade Agreement was estimated at 15 per cent by 1992 (Schlefer, 1992:115).

BIBLIOGRAPHY

• • • • • • • •

Books and Articles

Arcus, P.L. 1981. *Broilers and Eggs*. Ottawa: Economic Council of Canada, Report E/I.

Arnott, Margaret L. 1975. *Gastronomy: The Anthropology of Food and Food Habits*. The Hague: Mouton Publishers.

Arroyo, Gonzalo et al. 1981. "Transnational Corporations and Agriculture in Latin America." *LARU Studies*, Vol.IV, No.2.

Bannock, G., R.E. Baxter, and R. Rees. 1972. *The Penguin Dictionary of Economics*. 3rd ed. Markham, Ont.: Penguin Books Canada.

Baran, Paul A. and Paul M. Sweezy. 1966. *Monopoly Capital: An Essay on the American Economic and Social Order*. New York: Modern Reader Paperbacks.

Basran, G.S. and David A. Hay, eds. 1988a. *The Political Economy of Agriculture in Western Canada*. Toronto: Garamond Press.

Basran, G.S. and David A. Hay. 1988b. "Crisis in Agriculture in Western Canada: A Theoretical Explanation." In Basran and Hay (1988a).

Baxter, D. 1969. "The Canning Industry in Prince Edward County." Mimeo.

Bertin, Oliver. 1990. "Canada Packers Gets a Prime Steer." *The Globe and Mail*. May 4.

Bertin, Oliver. 1992. "Cargill Set to Invade Ontario Meat Market." *The Globe and Mail*. August 20.

Biggs, Everett. 1990. *The Challenge of Achievement: The Ontario Milk Marketing Board's First 25 Years of Operation, 1965 to 1990*. Mississauga, Ont.: The Ontario Milk Marketing Board.

Bitton, Joseph. 1985. "The Development and Merchandising of Generic Food Products: Implications of Pricing and Quality." M.Sc. thesis, McGill University, Montreal.

Blishen, Bernard R., Frank E. Jones, Kaspar D. Naegele, and John Porter, eds. 1964. *Canadian Society: Sociological Perspectives*. Revised edition. Toronto: Macmillan.

Bollman, Ray D. and Pamela Smith. 1988. "Integration of Canadian Farm and Off-Farm Markets and the Off-Farm Work of Farm Women, Men, and Children." In Basran and Hay (1988a).

Britnell, George E. 1939. *The Wheat Economy.* Toronto: University of Toronto Press.

Britnell, G.E. and V.C. Fowke. 1962. *Canadian Agriculture in War and Peace, 1935-50.* Stanford, Cal.: Stanford University Press.

Broadfoot, Dave. 1973. *Ten Lost Years, 1929-1939: Memories of Canadians Who Survived the Depression.* Toronto: Doubleday Canada.

Brodie, Janine. 1990. *The Political Economy of Canadian Regionalism.* Toronto: Harcourt Brace Jovanovich.

Burck, Charles G. 1979. "Plain Labels Challenge the Supermarket Establishment." *Fortune.* March 26.

Burman, Patrick. 1988. *Killing Time, Losing Ground: Experiences of Unemployment.* Toronto: Thompson Educational Publishing.

Burns, Jim, ed. 1983. *The Food Industry: Economics and Policies.* London: Heinemann.

Buttel, Frederick H. and Howard Newby, eds. 1980. *The Rural Sociology of the Advanced Societies.* Montclair, N.J.: Allenheld, Osmun.

Buzzell, Robert, John Quelch, and Walter Salmon. 1990. "The Costly Bargain of Trade Promotion." *Harvard Business Review,* March-April.

Canada. 1931. *Census of Merchandising and Service Establishments "Food Chains."* Ottawa: Dominion Bureau of Statistics.

Canada. 1933. *Census of Merchandising and Service Establishments "Food Chains in Canada 1930."* Ottawa: Dominion Bureau of Statistics.

Canada. 1937. *Report of the Royal Commission on Price Spreads.* Ottawa.

Canada. 1944. *Census of Merchandising and Service Establishments "Food Chains in Canada Calendar Year 1941."* Ottawa: Dominion Bureau of Statistics.

Canada. 1966, 1986. *Dairy Products Industries.* No. 32-209. Ottawa: Statistics Canada.

Canada. 1969. *Report of the Federal Task Force on Agriculture.* Ottawa.

Canada. 1971. *Report of the Royal Commission on Farm Machinery.* Ottawa.

Canada. 1978. *Report of the Royal Commission on Corporate Concentration.* Ottawa.

Canada. 1979. *Fruit and Vegetable Processing Industries: Annual Census of Manufacturers.* Ottawa: Statistics Canada.

Canada. 1981. *Canada's Agri-food System.* Ottawa: Agriculture Canada.

Canada. 1987. *The Daily: Census of Agriculture: 1986.* Ottawa: Statistics Canada.

Canada Packers. 1991. *Annual Report.*

Carlson, Jerry. 1971. "Big Corporations Back Out of Farming." *Farm Journal,* Vol.95, No.4 (April).

Clement, Wallace. 1975. *The Canadian Corporate Elite: An Analysis of Economic Power.* Toronto: McClelland and Stewart.

Clement, Wallace. 1983. *Class, Power and Property: Essays on Canadian Society.* Toronto: Methuen.

Clement, Wallace. 1984. "Canada's Coastal Fisheries: Formation of Unions, Cooperatives, and Associations." *Journal of Canadian Studies,* Vol.19, No.1.

Connor, John M. 1981. "Food Product Proliferation: A Market Structure Analysis." *American Journal of Agricultural Economics*, Vol.63 (November).

Connor, John M., Richard T. Rogers, Bruce W. Marion, and Willard E. Mueller. 1985. *The Food Manufacturing Industries*. Toronto: Lexington Books.

Cooperstock, Henry. 1968. "Prior Socialization and Cooperative Farming." In *Canadian Society: Sociological Perspectives*, 3rd ed., ed. Bernard R. Blishen, Frank E. Jones, Kasper Naegele, and John Porter. Toronto: Macmillan of Canada.

Corcoran, Terence. 1991. "Helping Canada's Food Business to Death." *The Globe and Mail*, April 7.

Corditz, Dan. 1978. "Corporate Farming: A Tough Row to Hoe." In *Change in Rural America: Causes, Consequences and Alternatives*, ed. R. Rodefeld et al. St. Louis: C.V. Mosby Company.

Cosman, Madeleine Pelner. 1976. *Fabulous Feasts: Medieval Cookery and Ceremony*. New York: George Braziller.

Cotterill, Ronald. 1984. "Market Structure-Price Relationships in Vermont Food Retailing Markets." *Working Paper*, University of Wisconsin, Madison: NC 117.

Cotterill, Ronald W. and Willard F. Mueller. 1979. "The Impact of Firm Conglomeration on Market Structure: Evidence for the U.S. Food Retailing Industry." *Working Paper 33*, University of Wisconsin, Madison: NC 117.

Crowley, Terry. 1990. *Agnes Macphail and the Politics of Equality*. Toronto: James Lorimer and Company.

Davies, Charles. 1987. *Bread Men: How the Westons Built an International Empire*. Toronto: Key Porter Books.

Davis, John E. 1980. "Capitalist Agricultural Development and the Exploitation of the Propertied Labourer." In Buttel and Newby (1980).

DiManno, Rosie. 1991. "Metro Poor Go Hungry Amidst Piles of Food." *The Toronto Star*, March 26.

Dominke, Arthur. 1976. *La Agroindustria en Mexico*. Mexico City: CIDE.

Duffy, Ann and Norene Pupo. *Part-Time Paradox: Connecting Gender, Work and Family*. Toronto: McClelland & Stewart, 1992.

Eatwell, John, Murray Milgate, and Peter Newman. 1987. "Invisible Hand." In *The New Palgrave: A Dictionary of Economics*. New York: Macmillan.

Elder, Louise. 1986. "The History of Canadian Canners, 1903-1986." Burlington, Ont.: mimeo.

Engelmann, Frederick C. 1954. "The Cooperative Commonwealth Federation of Canada: A Study of Membership Participation in Party Policy-Making." Ph.D. thesis, Yale University, New Haven, Conn.

Evans, Mark. 1990. "British Firm Bids to Build a New Canada Packers." *Financial Post*, April 18.

Fairbairn, Garry Lawrence. 1984. *From Prairie Roots: The Remarkable Story of the Saskatchewan Wheat Pool*. Saskatoon, Sask.: Western Producer Prairie Books.

Farb, Peter and George Armelagos. 1980. *Consuming Passions: The Anthropology of Eating*. Boston: Houghton Mifflin Company.

Fenton, Alexander and Eszter Kisban, eds. 1986. *Food in Change: Eating Habits from the Middle Ages to the Present Day*. Edinburgh: John Donald Publishers.

Ferguson, Ralph. 1991. *Compare the Share Phase I: Canadian Farmers Need a Fair Share of the Consumer Food Dollar*. Ottawa: House of Commons.

Ferguson, Ralph. 1992. *Compare the Share Phase II: The Comparisons Continue*. Ottawa: House of Commons.

Fiber, Ben. 1986. "Loblaws Focuses on Own Products." *The Globe and Mail*, June 9.

Finkel, Alvin. 1979. *Business and Social Reform in the Thirties*. Toronto: James Lorimer and Company.

Fowke, Vernon. 1946. *Canadian Agricultural Policy: The Historical Pattern*. Toronto: University of Toronto Press.

Fowke, Vernon. 1957. *The National Policy and the Wheat Economy*. Toronto: University of Toronto Press.

Friedmann, H. 1978a. "World Market, State and Family Farm: Social Bases of Household Production in the Era of Wage Labour." *Comparative Studies in Society and History*, Vol.20, No.4.

Friedmann, H. 1978b. "Simple Commodity Production and Wage Labour in the American Plains." *The Journal of Peasant Studies*, Vol.6, No.1.

Friedmann, H. 1979. "Political Economy of Food: Class Politics and Geopolitics in the World Wheat Economy." Department of Sociology, University of Toronto.

Friedmann, H. 1980. "State Policy and World Commerce: The Case of Wheat, 1815 to Present." Department of Sociology, University of Toronto.

Frundt, Henry J. 1975. "American Agribusiness and U.S. Foreign Agricultural Policy." Unpublished Ph.D dissertation, Rutgers University, New Brunswick, N.J.

Frundt, Henry J. 1981. "The Forces and Relations of Food Processing in the United States." Paper presented to the annual meetings of the Rural Sociological Society, University of Guelph, Guelph, Ont.

Galbraith, John Kenneth. 1956. "Inequality in Agriculture: Problem and Program." First J.J. Morrison Memorial Lecture, Ontario Agricultural College, Guelph, November 16, 1956, mimeo.

Gallo, Anthony. 1981. "Food Advertising." *National Food Review*, Vol.13 (Winter).

Gertler, Michael Eden. 1991. "The Institutionalization of Grower-Processor Relations in the Vegetable Industries of Ontario and New York." In *Towards a New Political Economy of Agriculture*, ed. William Friedland, Lawrence Busch, Frederick Buttel, and Alan Rudy. Boulder, Col.: Westview Press.

Goldberg, R.A. 1957. *A Concept of Agribusiness*. Boston: Harvard University Press.

Goldenberg, Susan. 1984. *The Thomson Empire*. Toronto: Methuen.

Goldschmidt, Walter. 1947. *As You Sow*. New York: Harcourt, Brace.

Goldschmidt, Walter. 1978. "Large-Scale Farming and the Rural Social Structure." *Rural Sociology*, Vol.43, No.3.

Good, W.C. 1958. *Farmer Citizen: My Fifty Years in the Canadian Farmers' Movement.* Toronto: The Ryerson Press.

Goodman, David and Michael Redclift. 1985. "Capitalism, Petty Commodity Production and the Family Enterprise." *Sociologia Ruralis*, Vol.25, Nos.3-4.

Green, Gary P. 1985. "Large-Scale Farming and the Quality of Life in Rural Communities: Further Specification of the Goldschmidt Hypothesis." *Rural Sociology*, Vol.50, No.2.

Hall, Lana, Andrew Schmitz, and James Cothern. 1979. "Beef Wholesale-Retail Marketing Margins and Concentration." *Econometrica*, Vol.46.

Hamm, Larry G. 1981. "The Impact of Food Distribution Procurement Practices on Food System Structure and Coordination." *Working Paper 58*, University of Wisconsin, Madison: NC 117.

Hamm, Larry G. 1982. "Retailer-Manufacturer Relationships in the Food-Service: Some Observations from the U.S.A." *Working Paper 64*, University of Wisconsin, Madison, Wis.: NC 117.

Hampe, Edward C. Jr. and Merle Wittenberg. 1964. *The Lifeline of America: Development of the Food Industry.* New York: McGraw-Hill.

Harris, Craig K. and Jess Gilbert. 1982. "Large-Scale Farming and Rural Income and Goldschmidt's Agrarian Thesis." *Rural Sociology*, Vol.47, No.3.

Harris, Marvin. 1985. *Good to Eat: Riddles of Food and Culture.* New York: Simon and Schuster.

Hedley, Max. 1979. "Domestic Commodity Production: Small Farmers in Alberta." In *Challenging Anthropology: A Critical Introduction to Social and Cultural Anthropology*, ed. David A. Turner and Gavin A. Smith. Toronto: McGraw-Hill Ryerson.

Heffernan, William D. and Douglas Constance. 1992. "Concentration and the Food Industry." Paper prepared for the World Farmers' Congress, Quebec City.

Heisler, Nick. 1991. "Food Processing Closures in Ontario as of October 31, 1991." . Ottawa: House of Commons, mimeo.

Hennessey, S.G. 1965. *Report of the Ontario Milk Inquiry Committee.* Toronto: Government of Ontario.

Henning, John A. and H. Michael Mann. 1978. *Issues in Advertising: The Economics of Persuasion.* Ed. David Ruerck. Washington, D.C.: American Enterprise Institute.

Hilton, Rodney et al. 1978. *The Transition from Feudalism to Capitalism.* New York: Verso.

Horn, Michiel. 1980. *The League for Social Reconstruction: Intellectual Origins of the Democratic Left in Canada, 1930-1942.* Toronto: University of Toronto Press.

Horsman, Mathew. 1990. "Hillsdown Profits Rise as Firm Eyes Canada." *Financial Post*, March 15.

Howe, D. 1983. "The Food Distribution Sector." In Burns (1983).

Irvine, William. 1976. *The Farmers in Politics, 1840-1911.* Toronto: McClelland and Stewart.

Janigan, Mary. 1990. "Why Chickens Don't Come Cheap." *The Globe and Mail Report on Business Magazine,* October 1990.

de Janvry, Alain. 1980. "Social Differentiation in Agriculture and the Ideology of Neopopulism." In Buttel and Newby (1980).

Johnston, Charles. 1986. *E.C. Drury: Agrarian Idealist.* Toronto: University of Toronto Press.

Johnston, Charles M. 1988. "Drury, Ernest Charles." In *The Canadian Encyclopedia,* 2nd ed. Vol I. Edmonton, Alta.: Hurtig Publishers.

Kneen, Brewster. 1990. *Trading Up: How Cargill, the World's Largest Grain Company, Is Changing Canadian Agriculture.* Toronto: NC Press.

Knuttila, Murray. 1989. "E.A. Partridge: The Farmer's Intellectual." *Prairie Forum,* Vol.14, No.1 (Spring).

Koller, Roland H. 1979. "Predatory Pricing: The Tip of the Iceberg." Paper presented at the American Economic Association annual meeting, Atlanta, Ga.

Korsching, Peter F. and Curtis W. Stofferahn. 1986. "Agriculture and Rural Community Interdependencies." In *Agricultural Change: Consequences for Southern Farms and Rural Communities,* ed. Joseph Molnar. Boulder, Col.: Westview Press.

Krause, Kenneth R. 1987. *Corporate Farming, 1969-82.* Economic Research Service Report, No. 578. Washington, D.C.: U.S. Department of Agriculture.

Lamm, McFall. 1981. "Prices and Concentration in the Food Retailing Industry." *Journal of Industrial Economics,* Vol.30.

Laycock, David H. 1990. *Populism and Democratic Thought in the Canadian Prairies, 1910 to 1945.* Toronto: University of Toronto Press.

Lipset, Seymour M. 1968. *Agrarian Socialism: The Cooperative Commonwealth Federation in Saskatchewan.* Garden City, N.Y.: Doubleday and Company.

Lobao, Linda. 1990. *Locality and Inequality.* Albany, N.Y.: State University of New York Press.

Lobao, Linda. 1987. "Farm Structure, Industry Structure, and Socio-Economic Conditions in the United States." *Rural Sociology,* Vol. 52, No. 4.

Lockyer, Peter. 1983. "An Uncertain Harvest: Hard Work, Big Business and Changing Times in Prince Edward County, Ontario." M.A. thesis, Carleton University, Ottawa.

Loubier, Yvan. 1984. *Structure and Economic Importance of the Canadian Food and Beverage Manufacturing Sector: Highlights of 1970-81.* Ottawa: Agriculture Canada.

Mackintosh, W.A. 1924. *Agricultural Cooperation in Western Canada.* Toronto: Ryerson Press.

Macpherson, C.B. 1975. *Democracy in Alberta: Social Credit and the Party System.* Toronto: University of Toronto Press.

MacPherson, Ian. 1979. *The Cooperative Movement on the Prairies, 1900-1955.* Booklet No.33. Ottawa: Canadian Historical Association.

Malassis, Louis. 1973. *Economie agro-alimentaire*. Vol.1. Paris: Editions Cujas.

Malassis, Louis and Martin Padilla. 1986. *Economie agro-alimentaire*. Vol.3. Paris: Editions Cujas.

Manchester, Allen. 1983. "The Farm and Food System: Major Characteristics and Trends." In *The Farm and Food System in Transition*. Economic Research Service, U.S. Dept. of Agriculture.

Maple Leaf Foods, Inc. 1991. *Annual Report*.

Marchak, M. Patricia. 1991. *The Integrated Circus: The New Right and the Restructuring of Global Markets*. Montreal and Kingston, Ont.: McGill-Queen's University Press.

Marfels, Christian. 1988. "Aggregate Concentration in International Perspective: Canada, Federal Republic of Germany, Japan, and the United States." In *Mergers, Corporate Concentration and Power in Canada*, ed. R.S. Khemani, D.M. Shapiro, and W.T. Stanbury. Halifax: The Institute for Research on Public Policy/ L'Institut de recherches politiques.

Marion, Bruce W. et al. 1979. *The Food Retailing Industry: Market Structure, Profits, and Prices*. New York: Praeger.

Marion, Bruce W. 1984. "Strategic Groups, Entry Barriers and Competitive Behaviour in Grocery Retailing." *Working Paper 81*, University of Wisconsin, Madison: NC 117.

Marion, Bruce W. 1986. *The Organization and Performance of the U.S. Food System*. Toronto: Lexington Books.

Marx, Karl. 1977. *Capital*. Vol.I. Moscow: Progress Publishers.

Matas, Robert. 1987. "Stocking Shelves Has a Hidden Cost." *The Globe and Mail*, February 28.

Matthie, K. 1982. "Regulated Marketing: The National View." *Agrologist*, Vol.11, No.4.

Matkin, James. 1990. "Time to Reform Supply System in Agriculture." *The Globe and Mail*, November 12.

McCormack, A. Ross. 1977. *Reformers, Rebels, and Revolutionaries: The Western Canadian Radical Movement, 1899-1919*. Toronto: University of Toronto Press.

McCormick, Veronica. 1968. *A Hundred Years in the Dairy Industry: A History of the Dairy Industry in Canada and the Events That Influenced It, 1867-1967*. Ottawa: Dominion Loose Leaf Co.

McNaught, Kenneth. 1959. *A Prophet in Politics: A Biography of J.S. Woodsworth*. Toronto: University of Toronto Press.

Mehri, Hussein. 1984. "The Development of Agribusiness in Quebec and Its Impact on the Family Farm: The Case of Poultry Production." M.A. thesis, Concordia University, Montreal.

Mitchell, Don. 1975. *The Politics of Food*. Toronto: James Lorimer and Company.

Morton, W.L. 1950. *The Progressive Party in Canada*. Toronto: University of Toronto Press.

Naylor, C. David. 1986. *Private Practice, Public Payment: Canadian Medicine and the Politics of Health Insurance, 1911-1966*. Montreal and Kingston, Ont.: McGill-Queen's University Press.

Nielsen Media Services. 1991. *National Advertising Expenditures Annual Summary 1991*. Toronto: Nielsen Marketing Research.

Niosi, Jorge. 1981. *Canadian Capitalism: A Study of Power in the Canadian Business Establishment*. Toronto: James Lorimer and Company.

Nova Scotia Fruit Growers Association (NSFGA). 1985. *Nova Scotia Tree Fruit Industry: Profile of the Future*. Ottawa: Agriculture Canada.

Organization for Economic Cooperation and Development (OECD). 1979. *Impact of Multinational Enterprises on National Scientific and Technical Capacities: Food Industry*. Paris.

OECD. 1988. *National Policies and Agricultural Trade: Country Study Sweden*. Paris.

Ontario. 1972. *Corporate Farming and Vertical Integration in Ontario*. Toronto: Ministry of Agriculture and Food, Economics Branch.

Ontario. 1986. *Agricultural Statistics for Ontario*. Publication 20, table 9. Toronto: Ministry of Agriculture and Food.

Ontario Milk Marketing Board. 1984. *Industrial Milk Allocation in Ontario: A Report Prepared for the Ontario Milk Marketing Board*. Mississauga, Ont.

Ontario Vegetable Growers Marketing Board. 1986a. *Agreement and Award for Marketing the 1986 Crop of Green Peas for Processing*. London, Ont.

Ontario Vegetable Growers Marketing Board. 1986b. *Agreement for Marketing the 1986 Crop of Green and Wax Beans for Processing*. London, Ont.

Park, Libbie and Frank Park. 1973. *Anatomy of Big Business*. Toronto: James Lewis & Samuel.

Partridge, E.A. 1926. *A War on Poverty*. Winnipeg: Wallingford Press.

Perkin, G.F. 1962. *Marketing Milestones in Ontario 1935-1960*. Toronto: Ontario Department of Agriculture.

Persigehl, Elmer and James York. 1979. "Substantial Productivity Gains in the Fluid Milk Industry." *Monthly Labor Review*, July.

Pigg, Susan. 1989. "Big Business Stocks U.S. Food Banks as 'Industry Builds around Poverty.'" *The Toronto Star*, November 2.

Pirie, Madsen. 1989. "The Principles and Practice of Privatization: The British Experience." In *Political Ideologies and Political Philosophies*, ed. H.B. McCullough. Toronto: Wall and Thompson.

Porter, Michael D. 1980. *Competitive Strategy*. New York: The Free Press.

Powell, Horace B. 1956. *The Original Has This Signature: W.K. Kellogg*. Englewood Cliffs, N.J.: Prentice-Hall.

Prescott, D.M. 1981. *The Role of Marketing Boards in the Process Tomato and Asparagus Industries*. Ottawa: Economic Council of Canada.

Raup, Philip. 1973. "Corporate Farming in the United States." *Journal of Economic History*, Vol.33.

Reguin, Eric. 1990. "Cargill Stalks Packers as Buyers Circle." *Financial Post*, March 10/12.

Resnick, A. and B. Stern. 1977. "An Analysis of Information Content in Television Advertising." *Journal of Marketing*, Vol.41.

Resnick, Philip. 1989. "The Ideology of Neo-Conservatism." In *Political Ideologies and Political Philosophies*, ed. H.B. McCullough. Toronto: Wall and Thompson.

Rosenbluth, Gideon. 1957. *Concentration in Canadian Manufacturing Industries*. Princeton, N.J.: Princeton University Press.

Scherer, Frederic M. 1982. "The Breakfast Cereal Industry." In *The Structure of American Industry*, 6th ed., ed. Walter Adams. New York: Macmillan Publishing.

Schlefer, Jonathan. 1992. "What Price Economic Growth?" *The Atlantic Monthly*, December.

Schmalensee, Richard. 1978. "Entry Deterrence in the Ready-to-Eat Breakfast Cereal Industry." *Bell Journal of Economics*, Vol.9.

Scott, C.D. 1984. "Transnational Corporations and Asymmetries in the Latin American Food System." *Bulletin of Latin American Research*, Vol.3, No.1.

Shapiro, Eben. 1992. "P&G Takes on the Supermarkets with Uniform Pricing." *The New York Times*, April 26.

Sharp, Paul. 1971. *The Agrarian Revolt in Western Canada*. New York: Octagon Books.

Silverstein, Sanford. 1968. "The Rise, Ascendancy and Decline of the Cooperative Commonwealth Federation Party of Saskatchewan, Canada." Ph.D. thesis, Washington University, St. Louis, Mo.

Stanbury, W.T. 1988. "Corporate Power and Political Influence." In *Mergers, Corporate Concentration and Power in Canada*, ed. R.S. Khemani, D.M. Shapiro, and W.T. Stanbury. Halifax: Institute for Research on Public Policy/L'Institut de recherches politiques.

Stirling, Bob and John Conway. 1988. "Fractions Among Prairie Farmers." In Basran and Hay (1988a).

Swanson, Louis. 1988. *Agriculture and Community Change in the U.S.* Boulder, Col.: Westview Press.

Taylor, Malcolm G. 1978. *Health Insurance and Canadian Public Policy: The Decisions that Created the Canadian Health Insurance System*. Montreal: McGill-Queen's University Press.

Teuteberg, Hans J. 1986. "Periods and Turning-Points in the History of European Diet: A Preliminary Outline of Problems and Methods." In Fenton and Kisban (1986).

Traves, Tom. 1979. *The State and Enterprise: Canadian Manufacturers and the Federal Government, 1917-1931*. Toronto: University of Toronto Press.

Valaskakis, Kimon. 1992. "A Prescription for Canada Inc." *The Globe and Mail*, October 31.

Veeman, T.S. and M.M. Veeman. 1978. "The Changing Organization, Structure and Control of Canadian Agriculture." *American Journal of Agricultural Economics*, Vol.60, No.5.

Veltmeyer, Henry. 1987. *Canadian Corporate Power*. Toronto: Garamond Press.

Warnock, John W. 1978. *Profit Hungry: The Food Industry in Canada*. Vancouver: New Star Books.

Warnock, John W. 1988. *Free Trade and the New Right Agenda*. Vancouver: New Star Books.

Warnock Hersey International Ltd. 1970. *Fruit and Vegetable Processing Industry: Atlantic Provinces*. Ottawa: Department of Regional Economic Expansion.

Wessel, James. 1983. *Trading the Future*. San Francisco: Institute for Food and Development Policy.

Willis, J.S. *This Packing Business: The History and Development of the Use of Meat to Feed Mankind, From the Dawn of History to the Present*. Toronto: Canada Packers, 1964.

Wilson, C.F. 1978. *A Century of Canadian Grain: Government Policy to 1951*. Saskatoon, Sask.: Modern Press.

Winson, Anthony. 1985. "The Uneven Development of Canadian Agriculture: Farming in the Maritimes and Ontario." *The Canadian Journal of Sociology*, Vol.10, No.4.

Winson, Anthony. 1988. "Researching the Food Processing-Farming Chain: The Case of Nova Scotia." *The Canadian Review of Sociology and Anthropology*, Vol.25, No.4.

Winson, Anthony. 1990. "Capitalist Coordination of Agriculture: Food Processing Firms and Farming in Central Canada." *Rural Sociology*, Vol.55, No.3.

Wood, Louis Aubrey. 1975. *A History of Farmers' Movements in Canada*. Toronto: University of Toronto Press.

World Bank. 1988. *World Development Report 1988*. New York: Oxford University Press.

Wygant, P. 1978. *Report of the Task Force on the Food and Beverage Industry*. Ottawa.

Young, Walter. 1969. *The Anatomy of a Party: The National CCF, 1932-1961*. Toronto: University of Toronto Press.

Young, Walter. 1969. *Democracy and Discontent: Progressivism, Socialism and Social Credit in the Canadian West*. Toronto: Ryerson Press.

Zakuta, Leo. 1964. *A Protest Movement Becalmed: A Study of Change in the CCF.* Toronto: University of Toronto Press.

Newspapers and Journals

Canadian Business Magazine. November 1990.

Canadian Grocer. January 1990; May, June, August 1991.

Consumer Report. February 1981.

The Edmonton Journal. May 26, 1987.

The Farmer's Magazine. September 1952.

Farmers' Sun. 1922-23.

Financial Post. Various issues; Yellow Cards.

The Gazette (Montreal). Various issues.

The Globe and Mail. Various issues.

The Globe and Mail Report on Business Magazine. July 1989, October 1990.

The Ontario Milk Producer. 1932-36.

PDR Notes. Ontario Ministry of Agriculture and Food. Various issues, 1987-92.

The Toronto Star. Various issues.

INDEX

• • • • • • • •